Allocating Federal Funds for Science and Technology

Committee on Criteria for Federal Support of Research and Development

NATIONAL ACADEMY OF SCIENCES
NATIONAL ACADEMY OF ENGINEERING
INSTITUTE OF MEDICINE
NATIONAL RESEARCH COUNCIL

NATIONAL ACADEMY PRESS
Washington, D.C. 1995

NOTICE: The project that is the subject of this report was approved by the Governing Board of the National Research Council, whose members are drawn from the councils of the National Academy of Sciences, the National Academy of Engineering, and the Institute of Medicine. The members of the committee responsible for the report were chosen for their special competences and with regard for appropriate balance.

This report has been reviewed by a group other than the authors according to procedures approved by a Report Review Committee consisting of members of the National Academy of Sciences, the National Academy of Engineering, and the Institute of Medicine.

The National Academy of Sciences is a private, nonprofit, self-perpetuating society of distinguished scholars engaged in scientific and engineering research, dedicated to the furtherance of science and technology and to their use for the general welfare. Upon the authority of the charter granted to it by the Congress in 1863, the Academy has a mandate that requires it to advise the federal government on scientific and technical matters. Dr. Bruce Alberts is president of the National Academy of Sciences.

The National Academy of Engineering was established in 1964, under the charter of the National Academy of Sciences, as a parallel organization of outstanding engineers. It is autonomous in its administration and in the selection of its members, sharing with the National Academy of Sciences the responsibility for advising the federal government. The National Academy of Engineering also sponsors engineering programs aimed at meeting national needs, encourages education and research, and recognizes the superior achievements of engineers. Dr. Harold Liebowitz is president of the National Academy of Engineering.

The Institute of Medicine was established in 1970 by the National Academy of Sciences to secure the services of eminent members of appropriate professions in the examination of policy matters pertaining to the health of the public. The Institute acts under the responsibility given to the National Academy of Sciences by its congressional charter to be an adviser to the federal government and, upon its own initiative, to identify issues of medical care, research, and education. Dr. Kenneth I. Shine is president of the Institute of Medicine.

The National Research Council was established by the National Academy of Sciences in 1916 to associate the broad community of science and technology with the Academy's purposes of furthering knowledge and advising the federal government. Functioning in accordance with general policies determined by the Academy, the Council has become the principal operating agency of both the National Academy of Sciences and the National Academy of Engineering in providing services to the government, the public, and the scientific and engineering communities. The Council is administered jointly by both Academies and the Institute of Medicine. Dr. Bruce Alberts and Dr. Harold Liebowitz are chairman and vice chairman, respectively, of the National Research Council.

Support for this project was provided by the Department of Defense (under Contract No. N00014-95-C-0314), the National Institutes of Health (under Contract No. N01-OD-4-2139, Task Order #4), the National Science Foundation (under Grant No. OPS-9528889), and the Basic Science Fund of the National Academy of Sciences. Any opinions, findings, and conclusions or recommendations expressed in this material are those of the authors and do not necessarily reflect the views of the National Science Foundation.

Library of Congress Catalog Card Number 95-71602
International Standard Book Number 0-309-05347-1

Additional copies of this report are available from:

National Academy Press
2101 Constitution Avenue, NW
Box 285
Washington, DC 20055
800-624-6242
202-334-3313 (in the Washington Metropolitan Area)

B-680

Copyright 1995 by the National Academy of Sciences. All rights reserved.
Printed in the United States of America

Committee on Criteria for Federal Support of Research and Development

FRANK PRESS, Carnegie Institution of Washington, *Chair*
LEW ALLEN, JR., Charles Stark Draper Laboratory, Inc.
DAVID H. AUSTON, Rice University
FOREST BASKETT, Silicon Graphics Computer Systems
BARRY R. BLOOM, Albert Einstein College of Medicine
DANIEL J. EVANS, Daniel J. Evans & Associates
BARUCH FISCHHOFF, Carnegie Mellon University
MARYE ANNE FOX, University of Texas at Austin
SHIRLEY A. JACKSON, U.S. Nuclear Regulatory Commission[1]
ROBERT I. LEVY, Wyeth-Ayerst Research[2]
RICHARD J. MAHONEY, Monsanto Company (*retired*)
STEVEN L. McKNIGHT, Tularik, Inc.
MARCIA K. McNUTT, Massachusetts Institute of Technology
PAUL M. ROMER, University of California at Berkeley
LUIS SEQUEIRA, University of Wisconsin
HAROLD T. SHAPIRO, Princeton University
H. GUYFORD STEVER, Trustee and Science Advisor
JOHN P. WHITE, Department of Defense[3]

National Research Council Staff and Consultants

Norman Metzger, Study Director
Robert M. Cook-Deegan, Senior Program Officer
Christopher T. Hill, George Mason University
Michael G.H. McGeary, Consultant
Julie M. Esanu, Research Assistant
Danielle Dehmler, Project Assistant

---

[1] Resigned on July 12, 1995, to become chair of the U.S. Nuclear Regulatory Commission.
[2] Resigned on March 22, 1995, due to schedule conflicts.
[3] Resigned on June 22, 1995, to become deputy secretary of defense.

Preface

In a report accompanying funding for the National Institutes of Health for Fiscal Year 1995, the Senate Appropriations Committee requested a study from the National Academy of Sciences, the National Academy of Engineering, and the Institute of Medicine. The study was to address "the criteria that should be used in judging the appropriate allocation of funds to research and development activities, the appropriate balance among different types of institutions that conduct such research, and the means of assuring continued objectivity in the allocation process." The study originated from the Appropriations Committee's concern "that at a time when there is much opportunity to understand and cure disease, funding for health research supported by NIH in the next fiscal year is held to below the inflation rate for medical research due to budget constraints. Similarly, other Federal research agencies are confronted with constrained resources resulting from the virtual freeze in discretionary outlays."

The charge was daunting when it was requested by the Appropriations Committee and is even more so now. With a year's passage, the concern with a "virtual freeze in discretionary outlays" seems an understatement. The efforts by both the Administration and the Congress to reduce the federal deficit have prompted proposals to cut programs, consolidate or abolish agencies, and even do away with whole departments. The federal research and development enterprise has not been exempt from examination, nor should it be. Since the end of World War II, this enterprise has become vast and complex, and it accounts for a significant part of the discretionary outlays of the federal government. It is thus important that the nature and structure of federal support for research and development, as well as the benefits it brings, be understood to assure that as budgets are reduced, the strengths of U.S. science and technology are maintained, while the anachronistic or weak aspects are pruned.

The Committee on Criteria for Federal Support of Research and Development approached its task with realism about the budget pressures, an eagerness to provide advice that could guide both the Executive Branch and Congress, and a concern for fairness in evaluating the many parts of the enterprise. The committee's membership reflected these aims, including individuals who perform federally funded research, who use the results in industry and other sectors, who have been involved in shaping federal research and development programs in the past, and who are students of the research and development enterprise.

The committee's realism about budget pressures was matched by its realism about the report's immediate impact on current budgets. It is the committee's hope that this report will serve well both the executive and legislative branches as they grapple with the very hard decisions that will have to be made over many budget cycles, in a politically and fiscally difficult environment.

The theme of the committee's report is continuance in the face of change. *Continuance* builds on the spectacularly successful results of postwar federal investments in research and development. By any measure, these investments have been recouped many times over in contributing to a strong and globally competitive U.S. economy, hastening the end of the Cold War, providing continuing national security against new enemies, advancing the fight against disease, improving our environment, and producing revelations about ourselves, our world, and our cosmos. *Change* comes in acknowledging that the federal research and development enterprise must adapt to a new world. The Cold War is over. Global competition is both economic and military, involving many more nations than did the past bipolar confrontation of nuclear superpowers. These problems create opportunities. Indeed, science and technology will be even more important in the future than they are today. Change is also reflected in the very doing of science, as computers and high-speed communication networks expand access to databases and facilities throughout the world and enable daily collaboration among scientists and engineers separated by great distances.

Over time, institutions and programs have been created that no longer serve us well. Even good programs and institutions must give way to successors that are better and are more closely linked to new national needs. These are painful messages. Some of the committee's members have built their professional lives through programs and institutions that may not survive application of the principles the committee proposes for judging future expenditures. At the same time, the committee believes strongly that failure to make these choices will prove costly, serving neither the nation nor the scientific community. That said, the committee appreciates that its principles for judging programs and institutions are, by necessity, general and must be given more specificity when applied to particular programs and institutions. As a practical matter, the committee did not offer specific details for implementing the judgments that must be made. The committee believes that those who must make the decisions and execute them should be given the latitude to apply these principles sensibly.

The report is short, and deliberately so. Part I offers the committee's recommendations, with sufficient elaboration to enable readers to understand them. The four supplements included in Part II give details underlying the recommendations. These supplements are not mere appendixes, but provide background critical to understanding this brief report. For example, Supplement 2 shows how the committee derived a new budget index it calls federal science and technology (FS&T). The committee believes that these federal funds best define the public investment in the science and technology base that is essential for maintaining U.S. health, prosperity, and security.

In addition to the facts and analyses provided in the supplements, the committee relied on other means for arriving at its judgments, including more than 35 letters received from individuals in leadership positions in industry, academia, and scientific societies; a number of outreach meetings held around the country; several commissioned papers; communications through an Internet home page; briefings by senior government officials whose agencies are collectively responsible for most of the federal research and development budget; and discussions with many individuals in the Administration and Congress. The committee is grateful to all who took the time to provide assistance and in doing so not only tutored us, but also showed their concern for the future of the U.S. research and development enterprise. The individuals who assisted the committee and the background papers prepared for it are acknowledged in Appendixes C and D, respectively.

Some will think it politically unwise that we recommend a process and guidelines for identifying activities that can be reduced or eliminated and for reallocating the savings to ones more essential to preserving U.S. leadership in science and technology. We have been told that our advice will be only partially followed—that the cuts will be made but that the savings will not be reallocated to federal science and technology. Perhaps. But we see no alternative. We can only hope that the case we have made is convincing, and trust that our recommendations to maintain U.S. strength in science and technology will be accepted. The committee believes that the political wisdom that created the remarkably successful U.S. research and development enterprise will endure, driven by the U.S. public's strong and abiding support for federal science and technology.

This report results from the work of many people. I especially thank the committee itself. It had what some believed a near-impossible task. Whether it succeeded is for others to judge. I shall always be grateful to these extraordinarily accomplished and able people for the care, intelligence, and above all the time they gave to wise and experienced judgments about a federal role that is so vital to the nation's future. Finally, I know I speak for all the committee members in acknowledging our indebtedness to the staff—consummate professionals who know as much about science policy issues as any in Washington, and without whose participation the report would be much diminished.

                                            Frank Press
                                            Chair
                                            Committee on Criteria for Federal
                                            Support of Research and Development

Contents

Part I—Improving the Allocation Process for Federal Science and Technology

Determining Principles for Allocating Federal Funds	3
Conclusions, Recommendations, and Discussion	8
Looking to the Future	30
Endnotes	32

Part II—Supplements: Background and Rationale

1	The Evolution and Impact of Federal Government Support for R&D in Broad Outline	41
2	Federal Funds for R&D and FS&T	51
3	Current Processes for Allocating Federal R&D Funds	62
4	Interactions Between Federal and Industrial Funding and the Relationship Between Basic and Applied Research	70
Endnotes		82

Appendixes

A	Senate Report Language for the Prospective Study	87
B	Committee and Staff Biographical Information	88
C	Acknowledgments	93
D	List of Commissioned Background Papers	95
E	Acronyms	96

Part I

Improving the Allocation Process for Federal Science and Technology

Determining Principles for Allocating Federal Funds

The federal government has played a pivotal role in developing the world's most successful system of research and development. Over the past 5 decades the U.S. scientific and technical enterprise has expanded dramatically, and the federal investments in it have produced enormous benefits for the nation's economy, national defense, health, and social well-being. Science and technology will be at least as important for our nation's future as they have been for our past, but further expansion of federal funding for research and development is unrealistic in the next several years. Both the current administration's 10-year budget plan and the 7-year plans passed by the House and Senate propose significant reductions in federal discretionary spending. Maintaining the vigor of research and development is important—indeed essential—to the nation's future and will require the ability to increase funding for new opportunities selectively, even while reducing the overall budget.

The Committee on Criteria for Federal Support of Research and Development believes that it will be possible to sustain this country's scientific and technological preeminence and the strong federal role within current fiscal constraints if the recommendations in this report are adopted. Ensuring the nation's future health, however, may well require augmented investments later—after the current period of reorganization and consolidation has helped control costs and sharpen focus.

As we consider how to restructure federally funded research and development to meet today's budget realities, it is important to recognize the considerable strengths of the current system (see Supplement 1 for historical background). Those strengths should not be lost. "Top-down" mission-oriented management and "bottom-up" investigator-initiated research projects have combined to create a powerful research and development engine that is the envy of the world. Computer science, surface science, molecular biology, and other fields have emerged in response to new opportunities, and widely disparate fields have been combined to create entirely new applications. Competitively funded research and development projects subject to national merit review and conducted in every state of our nation have proven particularly effective. Federally funded university science and engineering, in addition to yielding new discoveries, has produced new generations of scientists and engineers who serve in academia, industry, and government and also fill critical management positions there. Investments in science have dramatically expanded our knowledge of ourselves and our universe, and new technologies have improved our daily lives. The fruits of federally funded research and development have been applied effectively by U.S. industry. Drawing on the support provided by many sponsoring agencies and the results from a wide range of performing institutions, the American entrepreneurial spirit has tapped federally funded research and development to form entirely new industries in areas such as microelectronics, biotechnology, and communications and information technology, among others.

The federal government invests in a portfolio of highly diversified activities in research and development in many disciplines—but there has not been an actively managed federal "budget." With the exception of selected recent initiatives, the federal R&D budget has been tallied up after the fact—it is the sum of R&D expenditures from federal departments and agencies used mainly for comparison with other federal expenditures or with the R&D budgets of other industrialized nations. Because it is added together after the individual budget and appropriations decisions have been made, it has never been "managed" as a coherent whole. Yet there *is* a federal process—one that engages a broad range of issues, complex interactions, and conflicts—from which de facto priorities emerge. Those priorities reflect contending goals, different performers (public or private; university, industry, or federal laboratories), multiple funding sources (almost every federal department and agency), competing jurisdictions (executive and legislative branches; budget, appropriations, and authorization committees within Congress), and international economic competition (proprietary national investment or international cooperation).

The extraordinary success of U.S. research and development can be continued within current budget constraints. However, ensuring continuing success will require rigorous discipline and a coherent and comprehensive approach for deciding how resources are used. This report proposes a new process for allocating and monitoring federal spending for science and technology across disciplines and government agencies. With an integrated view and a coherent federal science and technology budget, it will be possible to make selective reductions in some areas, so as to free badly needed resources for more productive investments and new opportunities that arise.

Defining a Federal Science and Technology Budget

To obtain advice on an appropriate budget design, Congress asked this committee to recommend criteria for federal support of research and development. Federal research and development expenditures are reported in current budget documents as being more than $70 billion annually.[1] Almost half of this amount, however, is spent on such activities as testing and evaluation of new aircraft and weapons systems in the Department of Defense, nuclear weapons work in the Department of Energy, and missions operations and evaluation in the National Aeronautics and Space Administration. Those activities are very important, but they involve the demonstration, testing, and evaluation of current knowledge and existing technologies. Even when they are technologically advanced, these functions do not involve the creation of new knowledge and the development of new technologies. The federal research and development budget as currently reported is thus misleading, because it includes large items that do not conform to the usual meaning of research and development.[2]

In studying how to ensure the continuing vibrance of U.S. research and development, the committee focused on the $35 billion to $40 billion in federal research and development spent annually on expanding fundamental knowledge and creating new technologies (see Supplement 2). Those are the expenditures that constitute federal support for a national science and technology base that underlies not only defense and space programs, but also the advancement of scientific knowledge and new technology used in many fields and industries. To focus discussion and more clearly identify this investment component of the federal research and development budget, the committee developed the term *federal science and technology* (FS&T) and an accompanying budget index (for details, see Supplement 2, especially Boxes II.3 and II.4). FS&T is used throughout this report to describe federal funding for those science and technology activities that produce or expand the use of new knowledge and new or enabling technologies (for examples, see Table I.1).

The committee recommends that, in the future, government support for basic and applied science and technology be presented, analyzed, and considered in terms of an FS&T budget. The current FS&T budget of $35 billion to $40 billion, including both training and research and development, represents about 0.5 percent of the nation's gross domestic product (see Box II.3 for background and definition). The distribution of funds for research and development as traditionally reported, compared to FS&T, illustrates the difference between the two concepts. Private industry performs the largest share of federally funded research and development as traditionally reported, but most of this work is downstream product demonstration, testing, and evaluation that is excluded from the committee's recommended new measure. When the FS&T measure is used instead, industry drops from first to third. Federal laboratories (both in-house and contractor-run) account for the largest share (39%) of FS&T, followed by academic institutions (31%), industry (21%), and nonprofit and other institutions (9%). (See Supplement 2 for additional details.)

The committee's definition of FS&T deliberately blurs any distinction between basic and applied science or between science and technology (see Table I.1). A complex relationship has evolved between basic and applied science and technology. In most instances, the linear sequential view of innovation is simplistic and misleading. Basic and applied science and technology are treated here as one interrelated enterprise, as they are conducted in the science and engineering schools of our universities and in federal laboratories. For further explanation of why the committee aggregates these activities within a single budget, see Supplements 1 and 4.

Structure and Approach of This Report

Part I of this report focuses on the committee's 13 recommendations for improving the process of allocating federal funds for science and technology. The conclusions, recommendations, and discussion are organized and presented to serve the following five purposes:

1. Make the allocation process more coherent, systematic, and comprehensive;
2. Determine total federal spending for federal science and technology, based on a clear commitment to ensuring U.S. leadership;
3. Allocate funds to the best projects and people;
4. Ensure that sound scientific and technical advice guides allocation decisions; and
5. Improve federal management of research and development activities.

Part II contains four supplements that provide critical background for and explain the rationale behind the committee's recommendations. Supplement 1 briefly surveys science policy and the impact of federal support since World War II; Supplement 2 describes the derivation of the FS&T budget number; Supplement 3 outlines the existing process for allocating funds; and Supplement 4 treats the distinction between basic and applied research and the interplay between federal and industrial funding. Four appendixes give details that bear on committee process and background. A fifth lists the acronyms used in this report.

TABLE I.1 Federal Science and Technology: Examples of Work That Enables Continuing U.S. Innovation

Characteristics	Examples (Funding Agencies)
Basic Research	
Creates new knowledge; is generic, non-appropriable, and openly available; is often done with no specific application in mind; requires a long-term commitment	Characterizing the mechanism of Alzheimer's disease—at many universities and NIH (NIH)
	Studying the physics of cloud formation—at universities and the National Center for Atmospheric Research (NOAA, NSF)
	Exploring the chemistry of photosynthesis—at many universities and federal laboratories (USDA, NSF)
	Elucidating basic components of matter through particle physics—at Fermi Laboratory and many universities (DOE, NSF)
	Understanding how earthquakes and volcanoes are related to plate tectonics—at universities and USGS laboratories (USGS, NSF)
	Exploring the changes in the universe over time through astronomy and cosmology—at universities, national laboratories, and NASA centers (NSF, NASA, DOE)
	Studying how language is acquired—at universities (NSF, NIH)
	Studying risk perception and methods of risk management—at universities and EPA, DOE, and DOD laboratories (EPA, DOE, DOD, NSF)
Applied Research	
Uses research methods to address questions with a specific purpose; pays explicit attention to producing knowledge relevant to producing a technology or service; overlaps extensively with basic research; can be short- or long-term	Predicting ground motion and landslides caused by earthquakes—at universities and federal laboratories (USGS)
	Discovering flexible, non-brittle, manufacturable, high-temperature superconducting wire—at Los Alamos National Laboratory and universities (DOE, DOD)
	Conducting clinical research on cancer chemotherapy and clinical trial methodology—at NIH, FDA, and academic health centers (NIH, FDA, CDC)
	Studying ethnography and sociology of drug abuse rituals related to AIDS transmission—at state health departments and universities (NIH, CDC)
	Studying econometric projection techniques—in universities and various federal agencies (NSF, DHHS, USDA, DOD)

TABLE I.1 Continued

Characteristics	Examples (Funding Agencies)

Applied Research (continued)

Discovering diagnostics and vaccines to combat emerging infections—at universities, foreign research centers, and CDC, NIH, and DOD laboratories (DOD, CDC, NIH, USAID)

Designing a new programming language—at universities and software companies (DOD, NSF)

Fundamental Technology Development

Develops prototypes; uses research findings to develop practical applications; is of general interest to a sector or sectors, but full returns cannot be captured by any one company; is usually short-term, but can be long-term; is not developed for one identifiable commercial or military product; often makes use of new knowledge from basic or applied research	Building an optical computer—at universities and computer firms (NSF, DOD) Developing new approaches to parallel processing, software, and hardware—at FFRDCs, universities, and private firms (DOD, NSF) Building a prototype DNA sequencing machine—at Caltech (NSF) Conducting clinical trials of a drug to treat heroin addiction—at VA hospitals, NIH, and academic health centers (DVA, NIH) Developing high-temperature ceramics for internal combustion engines—at universities and FFRDCs (NIST, DOD) Studying vitrification for storage of nuclear and hazardous waste—at national laboratories and some university engineering departments (DOE, EPA) Identifying a specific laser for use in guided missiles (before use in any one missile)—at DOD and university laboratories (DOD) Adapting cognitive science of language recognition for development of natural-language software—at universities and national laboratories (NSF, NIH, DOD) Developing strong, high-temperature alloys for engines, but not for a jet engine for a particular aircraft—at universities, NASA centers, DOD laboratories, and private firms (NASA, DOD) Breeding drought-resistant or saline-tolerant crop plants—at USDA centers and universities (USDA, USAID) Adapting fiber-optic laser surgery for prostate cancer—at universities and national laboratories (DOE, DOD, NIH) Developing a prototype for a walking robot—at FFRDCs, universities, and national laboratories (NASA, DOD, NSF, DOE)

Conclusions, Recommendations, and Discussion

The committee believes that the following 13 recommendations, as a set, will enable continuance of a strong federal research and development system at a time of change and stress.

The United States Must Develop a More Coherent Budget Process for Science and Technology.
(Recommendations 1-3)

RECOMMENDATION 1. The President should present an annual comprehensive FS&T budget, including areas of increased and reduced emphasis. The budget should be sufficient to serve national priorities and foster a world-class scientific and technical enterprise.

Currently, the federal research and development budget is typically defined as the sum of the research and development funds obligated or proposed by federal departments and agencies for programs and facilities classified as R&D. The research and development budget is never considered as an integrated whole during the development of the President's budget or given an overall review by Congress. Rather, the research and development budget is developed in the context of individual agency missions and programs.

Recent administrations have attempted to introduce more coherence in federal policy for R&D by creating an intergovernmental committee structure to coordinate budgeting for high-priority programs that involve more than one agency, for example, research on global change and on high-performance computing and communications.[3] The President may even single out certain programs or facilities as presidential initiatives. However, it has been difficult to shape those initiatives into integrated efforts that are more than an aggregation of agency programs that already exist. When the budget reaches Congress, it is disaggregated into the various appropriations bills and considered by many authorizing committees and appropriations subcommittees; efforts to achieve integrated initiatives can be quickly undone.

The existing approach works reasonably well during periods of growth, when new opportunities and shifts in emphasis can be accommodated within budget increases—without cutting back or closing down older activities that no longer rank as high priorities. But the disaggregated approach is less suitable when major cutbacks must be made. For example, the Department of Defense budget for research and development historically has supported the majority of federal funding for academic research and training in electrical engineering, metallurgy and materials, and computer science;[4] the Department of Energy is the largest contributor to other fields such as materials science (when national laboratories are included). All science and engineering depend critically on those fields, and cuts in Department of Defense and Department of Energy programs made for other purposes might well have significant and inadvertent impacts on diverse research and development programs conducted in many other agencies and having clear importance to the country. U.S. leadership in science and technology depends on more than the basic research supported by the National Science Foundation and the National Institutes of Health. It also depends on the science and engineering funded by the Department of Energy, Department of Defense, National Aeronautics and Space Administration, National Institute of Standards and Technology, and other mission agencies.

Budget cuts require an integrated consideration of their effects. Only in this way can the President and Congress determine the levels of investment for important, high-priority areas of research and development (especially those involving multiple agencies or reallocations among agencies), make the trade-offs needed to free up funds for new initiatives within the FS&T budget, and incorporate the results of systematic program and agency evaluations. Achieving such coordination will require significant changes in how the executive and legislative branches deal with the budget for federal science and technology. The requisite changes are discussed in Recommendations 2 and 3.

Questions to Consider in the Executive Office of the President

The President, the Office of Management and Budget, and the President's Science and Technology Advisor should employ a process that explicitly and publicly addresses pertinent questions, such as those listed below, as a means of providing budget guidance to agencies and a rationale to Congress and the public (see Box I.1 for a description of how the process might work).[5]

- Is the aggregate FS&T budget adequate to support the human and material resources that will maintain the United States as one of the leading nations in research and development in accord with the overarching national goals proposed in Recommendation 4 below?
- Does the FS&T budget recognize presidential initiatives, which might include national security needs; technical training of personnel in areas of national need; promising scientific opportunities; human spaceflight; research and development of economic importance, such as materials science; emerging public health problems; environmental or disaster mitigation; international projects; or responses to policies of other countries?
- Does the FS&T budget reflect overall federal budget constraints?
- Does the FS&T budget maintain strength by reallocating funds effectively?
- Are resources for laboratories, centers, and projects with obsolete missions or of insufficient quality being phased out, reduced, or redirected?
- Are measures proposed for reducing costs and inefficiencies?
- Is the FS&T budget appropriately balanced, and does it take account of the interdependencies of programs supported by different departments and agencies?

> **Box I.1**
> **PRIORITY SETTING AND DETERMINING FS&T BUDGETS AT THE PRESIDENTIAL LEVEL:**
> **HOW IT MIGHT WORK**
>
> At the beginning of the budget cycle, the President, with advice from the Director of the Office of Management and Budget and the President's Science and Technology Advisor,[1] decides on the aggregate level of funding for federal science and technology (FS&T) across the government that will maintain a leadership role for the United States and preserve the ability of agencies to perform their missions. Guidance is sent to agencies listing presidential priorities, including trade-offs and reallocations across agencies that reflect these priorities, as well as crises, opportunities, or evaluations. An extract of the President's budget message to Congress might read: "The federal science and technology budget is $XX billion dollars. Although this represents a reduction of $X billion, international comparisons show that it will enable us to maintain a world-class position in fundamental science and technology and a leadership position in the select fields of A, B, and C. The budget reduction was achieved by beginning to close and merge X federal laboratories and federally funded research and development centers (FFRDCs) as recommended by the laboratory-closing commission, and shutting down other programs no longer necessary or of poor quality. Within this budget reduction, I am recommending increases in funding for the physical sciences at the National Science Foundation; material sciences at federal laboratories, FFRDCs, and university materials research centers; research on the causes of violence at the National Science Foundation and on interventions to prevent it at the National Institute of Mental Health; research on genetic origins of disease at the National Institutes of Health; and microelectronics and sensor development in the Department of Defense programs. These initiatives will meet mission needs and contribute to the nation's overall strength in science and technology. . . . "
>
> ---
> [1] The Science and Technology Advisor has a variety of mechanisms to learn about opportunities to increase or decrease program budgets: the President's Committee of Advisors on Science and Technology, the National Science and Technology Council, and meetings with scientists and engineers from universities, federal laboratories, and industry, as well as meetings with science ministers from other countries.

RECOMMENDATION 2. Departments and agencies should make FS&T allocation decisions based on clearly articulated criteria that are congruent with those used by the Executive Office of the President and by Congress.

Examples of important questions to be considered by federal departments and agencies in allocating FS&T funding include the following (see Box I.2 for a description of how the process might work):

• Does the program under consideration contribute significantly to the agency's mission?
• Are there major new opportunities for research and development within the purview of this agency that should be proposed?

> **BOX I.2**
> **EVALUATION OF FS&T PROGRAMS AT THE DEPARTMENT AND AGENCY LEVEL: HOW IT MIGHT WORK**
>
> Cabinet secretaries or agency directors respond to presidential priorities and guidance. The National Science and Technology Council is a vehicle for coordinating cross-agency programs and assessing the adequacy of the entire FS&T budget. Budgets reflect federal fiscal realities, the results of performance evaluations, and the recommendations of special laboratory-review commissions, and they allow for trade-offs to support new opportunities and new missions by closing out projects and laboratories with outmoded missions or poor evaluations.
>
> A response to the President's stated priorities from the director of the National Institutes of Health and the secretary of Health and Human Services, for example, might look like the following:
>
> "Dear Mr. (or Ms.) President:
>
> "We recommend the termination of programs focused on A and the reduction of those focused on B, following an external review. The savings from those closings and reductions will total $XX million this year, but savings in future fiscal years will be larger, as shown in the accompanying projection. We propose to reallocate $X of those savings to high-priority items and emerging opportunities and problems. In response to your national priorities, we propose to increase funding for research by $X on the causes of violence and interventions to prevent it at the National Institute of Mental Health. In accord with your wishes to increase the national investment in the genetic origins of disease, $X million has been allocated, with $X going to the National Center for Human Genome Research, and the remainder going to several relevant institutes of the National Institutes of Health (NIH), as shown in the accompanying chart. . . .
>
> "Since the time of initial budget planning, we have become aware of the alarming spread of the "alpha" virus, a new infectious agent. The agent was identified by the rapid response of investigators in the NIH intramural research program, working with the Centers for Disease Control and Prevention in an international collaboration. We have used a fraction of the NIH discretionary account from the current fiscal year to fund small grant supplements to several academic health centers, as well as several laboratories in the intramural program of the NIH. Given the public health risk to the American people, we believe this is an urgent national priority, and NIH needs to mount a much larger and more permanent research program, including an extramural research effort to accompany our new intramural commitments. We request an additional $X million for this purpose. . . ."

- Does the allocation of budget reductions or increases recognize the highest-priority and highest-quality programs? Does it allow for new initiatives?
- Does the agency's external scientific and technical advisory body agree with the choices and priorities?
- Are the procedures for evaluating quality and mechanisms for using such evaluations both satisfactory?
- Does the peer or competitive merit review process used in recommended programs identify the best projects and performers, whether intramural or extramural? How is this demonstrated?

- Do programs recognize the importance of innovative and creative yet high-risk projects, interdisciplinary projects, and support for young scientists or engineers?
- Have trade-offs been made, cutting inferior or outmoded programs or divisions to reduce budgets and to enable new initiatives?
- Is the agency maintaining the infrastructure for research and development important to fulfilling its mission? Do decision makers recognize the importance of projects that both conduct research and train scientists and engineers?
- Does the allocation process fund the best performers equitably? Does it allow for the aspirations of institutions to improve their ability to compete and contribute nationally?
- Do reallocation decisions among classes of performers maintain a critical mass of expertise in federal agencies for effective priority setting, procurement, and public oversight?

RECOMMENDATION 3. Congress should create a process that examines the entire FS&T budget before the total federal budget is disaggregated into allocations to appropriations committees and subcommittees.

Decisions to allocate public funds are the prerogative of elected officials. The committee understands that members of Congress must address national needs but also represent the interests of constituents in their states or districts. In a time of severe fiscal constraints, public officials must decide among the many demands for government funds. The committee believes that the FS&T budget deserves special care because of its importance to the future of the country and because of the interdependence of its parts. Thus, the committee recommends that the FS&T budget be presented as a comprehensive whole in the President's budget and similarly considered as a whole at the beginning of the congressional budget process before the total federal budget is disaggregated and sent to the appropriations committees and subcommittees (see Box I.3 for a description of how the process might work). The committee recognizes that FS&T needs will be only one determinant of appropriations subcommittee allocations, but failure to take FS&T needs into account in advance risks harming the innovative enterprise that is key to the nation's future. Within the FS&T budget, it is crucial to be able to make trade-offs among agencies, programs, and performers in order to allow for new initiatives with funds freed by reducing or closing projects no longer needed or of insufficient quality.

The budget committees in both houses of Congress should take FS&T needs into account in the relevant budget function categories, such as defense, health, space, energy, agriculture, and general science. Budget resolutions do not determine appropriations decisions, however, but only set overall caps.[6] The appropriations committees therefore also must assess FS&T needs, both before and after deciding allocations to subcommittees, and when considering specific line items within agencies. Further, the subcommittees should consider research and develop-

> **Box I.3**
> **Considering and Evaluating a Comprehensive FS&T Budget in Congress: How It Might Work**
>
> The process of congressional evaluation begins with an assessment of the overall FS&T budget and the allocations to the departments and agencies. The chairs of the relevant authorization and appropriations committees are involved in a process that evaluates the proposed levels, trade-offs, reallocations, and cuts and increases across the government. The budget committees then assign funding levels to the several budget categories in which the FS&T budget is embedded. The Congressional Budget Office (CBO) tracks the FS&T pool as it is affected by the activities of the appropriations subcommittees and reports its status to the cognizant committee chairs. The committees and subcommittees undertake their process of hearings, consultations, and markups.
>
> One novel feature of this process is attention to the FS&T budget as a whole, and the trade-offs within it, before decisions are made about allocations to budget functions and to appropriations subcommittees. Another new feature is the monitoring of the FS&T pool throughout the process. Members, with the help of the CBO, can track the FS&T pool as trade-offs are made across and within agencies for the multiple purposes of meeting budget constraints; maintaining S&T leadership; fulfilling agency missions; responding to changing missions, opportunities, and crises; ensuring quality control and oversight; and accomplishing organizational reform.

ment needs and the FS&T budget as a whole as they allocate funds for agencies within their jurisdictions and make trade-offs against other spending.

A more coherent FS&T budget process in the Executive Branch should help Congress as well. The Carnegie Commission on Science, Technology, and Government recommended reorganization of the congressional committee structure and other measures.[7] Even without such reorganization, however, the current budget process could be improved by making it more open, soliciting better advice about research and development needs from outside experts, and assessing research and development needs early in the process. Recent administrations and Congresses have already taken steps in this direction, but further measures are needed.

Questions for Budget and Full Appropriations Committees to Consider

- Is the priority given to research and development adequate compared to the priority accorded other objectives in the government-wide discretionary budget?
- Is the total FS&T budget adequate to maintain a world-class level of scientific and technical performance by the United States?
- Does the President's FS&T budget sufficiently reflect fiscal constraints?
- Are the President's research and development priorities, trade-offs (e.g., reductions, closures, transfers, increases), and reallocations among agencies and programs appropriate?
- Are allocations to the various federal budget functions sufficient for agencies to perform their missions?

• Are there problems or opportunities identified by Congress that are not adequately accommodated in the President's FS&T budget?

Questions for Authorization Committees and Appropriations Subcommittees to Consider

• Do the priorities of the authorization committee or appropriations subcommittee agree with those of the budget committee?

• Does the authorization committee or appropriations subcommittee agree with the programs and allocations proposed for the agencies under its jurisdiction?

• Have the committees or subcommittees identified research areas, fields, or enabling technologies that are neglected or overfunded in the President's budget?

• Are items added to the FS&T budget by Congress intended to meet an important national need? Can the designated recipient institution make a national or regional contribution? Is the funding subject to external merit review? Has the item been aired in open hearings? Does it displace other FS&T investments of higher national priority?

• Will changes made by the committee or subcommittee have an impact on research and development programs outside its jurisdiction, and, if so, have they been taken into account?

Considering the FS&T budget as a coherent whole can improve the allocation process but cannot eliminate conflicts among agencies, among congressional committees and subcommittees, between the Senate and the House of Representatives, and between the executive and legislative branches. Such conflict is a part of the decentralized system of checks and balances in the U.S. federal system. The committee believes, however, that implementing Recommendations 1 through 3 will improve the budget process, better focusing the nation's public investment in research and development on the most important and promising opportunities.

The United States Should Strive to Continue as the World Leader in Science and Technology.
(Recommendations 4 and 5)

RECOMMENDATION 4. The President and Congress should ensure that the FS&T budget is sufficient to allow the United States to achieve preeminence in a select number of fields and to perform at a world-class level in the other major fields.[8]

The pool of approximately $35 billion to $40 billion in annual public support for FS&T is large and diverse. The committee believes that it is possible within that budget to reduce some programs, eliminate others, increase support of high-opportunity fields, and restrain federal spending—all while maintaining our nation's tradition of excellence in science and technology. To continue as a world leader,

> **Box I.4**
> **Evaluating FS&T Opportunities and Making International Comparisons:
> How It Might Work**
>
> Every five years, panels are convened to evaluate the fields in each major area of science and technology (e.g., physics, biology, electrical engineering), their standing in the world, and the resources needed to reach and maintain world-class position. Evaluation focuses on outputs, such as important discoveries, and also on certain benchmarks of best practice, such as number of scientists and engineers and their training or the current state of the laboratories and research facilities. To avoid conflicts of interest, at least half of the panel will include a few nonscientists plus experts from fields outside but related to the fields being evaluated. The panel will also include specialists in the evaluated fields who are recruited from the United States and foreign countries. If any field within a major area is performing below world standards but is judged to be a national priority, the panel will recommend that its budget be augmented or other changes made to bring it up to par. At the same time, the panel will identify the other fields with declining scientific opportunities and obsolete federal missions from which resources should be reallocated. Opportunities for international cost-sharing will be examined to achieve optimal use of federal funds devoted to science and technology.
>
> Evaluations will be commissioned by the National Science and Technology Council or its equivalent. The selection of fields for clear U.S. leadership from among those recommended by the panels will be made by the President and presidential advisors as part of the budget process. As an example, an extract of the President's budget message might read: "I propose that the United States need not be so far ahead in experimental particle physics, but should operate at world levels, in this case by contributing to construction of the particle accelerator in Geneva, sponsored by the CERN, and funding the participation of U.S. scientists in its design and research. On the advice of my Council of Advisors on Science and Technology, I propose that the United States should remain clearly preeminent in the molecular biology of plants and animals for the following reasons.... Accordingly, I will include the necessary additional funds in the FS&T budgets of the National Institutes of Health, the Department of Agriculture, and the National Science Foundation to achieve this goal...."

the United States should strive for clear leadership in the most promising areas of science and technology and those deemed most important to our national goals. In other major fields, the United States should perform on a par with other nations so that it is "poised to pounce" if future discoveries increase the importance of one of these fields. If the nation sets priorities in this way (see bulleted items below) and uses them in conjunction with the FS&T budget process, the result will be better decisions about reallocating and restructuring the U.S. research and development enterprise, preserving its core strengths, and positioning it well for strong future performance.

The international comparisons needed to assess U.S. achievement of its goals for leadership in research and development should be conducted by panels of the nation's leading experts under White House auspices. Reallocation decisions should be made with the advice and guidance of these expert panels, capable of determining the appropriate scope of the fields to assess and to judge the international stature of U.S. efforts in each field (see Box I.4 above for a discussion of how international comparisons might work). These panels would recommend to the President, his advisors, and Congress:

- Which fields must attain or maintain preeminence, based on goals such as economic importance, national security, unusual opportunity for significant discoveries, global resource or environmental issues, control of disease, mitigation of natural disasters, food production, a presidential initiative (such as human spaceflight), or an unanticipated crisis;
- Which fields require increases in funding, changes in direction, restructuring, or other actions to achieve these goals; and
- Which fields have excess capacity (e.g., are producing too many new investigators, have more laboratories or facilities than needed) relative to national needs and international benchmarks.

The committee believes that designing the budget process so as to secure an FS&T budget sufficient to ensure preeminence in select fields and world status in others will allow the United States to maintain continued world leadership. The FS&T budget process must be coupled to systematic review of investments by the nation's best scientific and technical experts, reporting to the highest reaches of government, to produce an appropriately balanced mix of activities. The committee emphasizes that wise federal investments will lead to the creation of new wealth in the future to an even greater extent than they have in the past. As a result, these investments will help reduce the federal deficit in the long run. After a period of budget constraints, reconfiguration, and adjustment, national needs may justify increased investments in FS&T.

RECOMMENDATION 5. The United States should pursue international cooperation to share costs, to tap into the world's best science and technology, and to meet national goals.

International cooperation is most clearly appropriate for large and expensive facilities such as high-energy accelerators and nuclear fusion facilities; for projects requiring coordinated research programs, such as many in oceanography as well as studies of global climate change; and for cross-national comparisons of health, education, and economic development.

Science is a global enterprise in which the United States must participate, for its own benefit and for that of the world. The scientific and engineering communities in the United States benefit from ideas and technologies developed all over the world; indeed, to remain world-class, the nation's scientists and engineers must be in touch with researchers around the globe. The United States also has important contributions to make in addressing the major problems of developing countries, such as disease, malnutrition, and overpopulation. In contributing to international scientific and technical collaborations and exchanges, enhancing free trade in ideas, and addressing major problems, the United States can contribute to improvements in the quality of life and pace of development in many countries. Ultimately, these efforts should also help expand global economic markets.

Maintaining U.S. Leadership in Science and Technology Despite Budget Constraints Will Require Discipline in the Allocation of Resources for Federal Investments. (Recommendations 6-9)

RECOMMENDATION 6. Research and development conducted in federal laboratories[9] should focus on the objectives of the sponsoring agency and not expand beyond the assigned missions of the laboratories. The size and activities of each laboratory should correspond to changes in mission requirements.

As described in Supplement 1, the present research and development system developed in the context of postwar economic expansion and the Cold War. Because the world has changed, we must reexamine the system of performers, phasing out weak or obsolete institutions (see Supplement 2, Box II.5, for a description of R&D performers).

Many reports on federal laboratories have been produced in recent years, including major reviews in the past year of Department of Defense, National Aeronautics and Space Administration, Department of Energy, and National Institutes of Health laboratories.[10] All conclude that federal laboratories have an important role in a balanced program of federal science and technology. Compared with extramural programs supporting academic and industrial research and development projects, federal laboratories offer distinctive features: relatively long term and stable funding of research programs; availability of unique facilities; full-time research opportunities without other distractions for staff scientists and engineers; closer links to the missions of their agencies; the ability to sustain programs for longer periods than those specified in the terms of a typical grant; and a capacity for rapid response to emergencies and sudden opportunities.[11] Many federal laboratories serve functions that, although they may not be at the frontiers of creating new knowledge, are nonetheless essential to science and technology, such as providing precise measurements and specification of standards or fulfilling specific program needs in health, defense, agriculture, the environment, forestry, and other areas.

Federal laboratories, however, have significant limitations. Study after study has shown the unfavorable environment that the federal government provides for research and development, through excessive and inflexible rules governing personnel, supplies, equipment, and facilities.[12] Today, federal laboratories also must accommodate shrinking budgets. Unfortunately, when government agencies receive fewer resources in real terms, the natural tendency is "to retain as much existing staff and infrastructure as possible in the face of a reduced budget, pull some contract work in-house, defer mission plans, and hope that future budgets will improve sufficiently ... to reinstate programs."[13] That tendency will be reinforced by pressures from local constituencies, because federal laboratories are major sources of employment and potential economic spin-offs. The committee believes that budget cuts provide a special impetus to a process that the federal laboratories should be following at all times: continual review of their success in meeting the missions of their agencies.

In the committee's view, some of the major reasons for supporting federal laboratories—both those the government operates directly and those operated by contractors—are less compelling than in the past. For some purposes, such as software system design and integration, private-sector firms increasingly have the highly sophisticated research and development capabilities that once justified unique arrangements with federally funded research and development centers (FFRDCs).[14] In addition, an increasing burden of federal regulations on those federal laboratories operated by universities and private firms has reduced many of the advantages of operation by nongovernment contractors, such as freedom from federal civil service restrictions and procurement regulations.

The damage and inefficiency induced by micromanagement from Washington emerge as major themes in the many reviews of federal laboratories.[15] Intrusions that come from agencies and through congressional mandates and earmarks are counterproductive, because any successful R&D laboratory must retain great flexibility and substantial autonomy to respond to rapidly paced scientific and technical change.

The end of the Cold War coupled with the pressures of the federal deficit have already affected the national laboratories and other FFRDCs significantly. National Science Foundation (NSF) reports estimate an 18 percent decrease for FFRDCs between 1992 and 1994,[16] and subsequent budget proposals by the President and Congress promise to cut substantially more. *There remain, however, superb FFRDCs that contribute uniquely to their agency's missions.*[17] It would be unwise to weaken these excellent performers. The recent review of NASA laboratories in fact pointed to several educational and management advantages of linking federally funded research to universities, and pointed to one NASA-funded FFRDC as a model for other NASA laboratories to emulate.[18] The general presumption, however, is against creating new federal laboratories when an alternative exists. Moreover, existing laboratories should undergo renewed scrutiny, with the possibility of redirecting or eliminating resources when mission requirements have diminished or if external reviewers judge that investments in a particular laboratory under review are less effective than other alternatives.

The February 1995 external review task force on the Department of Energy national laboratories concluded that they have clear expertise in their traditional mission areas of national security, energy, and environmental protection and in the fields of fundamental science underlying those missions (e.g., in basic research associated with high-energy, nuclear, and condensed-matter physics).[19] The task force viewed the DOE national laboratories as having "a distinctive role in conducting long-term, often high-risk R&D, frequently through the utilization of capital-intensive facilities which are beyond the financial reach of industry and academia, and generally through the application of multidisciplinary teams of scientists and engineers." However, the task force discouraged efforts of the DOE national laboratories to develop new missions, such as research and development in support of

U.S. industry and national competitiveness, arguing that other mechanisms were more effective or appropriate.[20] As a result, it concluded that the DOE national laboratory system should be "downsized" by refocusing on specific missions requirements, and it called for a more appropriate division of labor among the various performers—national laboratories, industrial research institutions, and research universities.[21]

The committee concurs with the general thrust of these recommendations. Federal laboratories should not seek new missions unless they offer both a critical advantage over other performers and the new mission better meets national needs. As with intramural laboratories, there is a natural tendency to maintain national laboratories and other FFRDCs with special relationships to their sponsoring agencies until the budget climate improves.[22] Their size and location make several DOE national laboratories particularly important sources of employment. Local factors are important to take into account in a transition strategy, but the size of the laboratories should in the long term be guided by mission requirements and national need. The best FFRDCs that serve the specialized needs of their sponsoring agencies should be sustained. Resizing of the national laboratory system should be balanced and appropriate within the larger division of labor among all federally funded performers of research and development.

The National Science and Technology Council (NSTC) recently produced a set of recommendations for NASA, DOE, and DOD laboratories.[23] NSTC noted "management problems that must be repaired" at NASA and DOE, particularly overstaffing within the agencies, overlap among missions of different laboratories, and excess micromanagement, especially at DOE. NSTC endorsed recent steps by NASA and DOE to reduce the size and simplify the management of their laboratories. NSTC judged management of DOD laboratories to be "generally effective," but noted that DOD "missed an opportunity" to improve cross-service integration, reduce redundancy, and shrink existing laboratories.[24]

The NSTC review and the agencies' own internal reviews, as well as the recent reviews of intramural research at NIH, are only now taking hold. The recommendations of the many reports, as well as oversight actions by Congress, should improve the effectiveness of the federal laboratory system, reducing its size and cost and improving its management. Federal laboratories will continue to play an important role in U.S. science and technology. The committee is concerned, however, that current reforms may bog down. The 1995 DOD review[25] recommended only a few major closings, for example. Recent reports on NIH, DOE, and NASA laboratories have not recommended closure of specific laboratories; however, the reports on NASA and DOE noted that such closures may be necessary in the future,[26] and a recent report on the largest NIH intramural program, the National Cancer Institute, recommended significant shrinkage.[27] If current initiatives do not achieve sufficient reductions, so that the federal laboratory system matches mission requirements, further steps may be necessary. Given the scale of the laboratories and their local economic significance, a device like the Base Closure and Realignment Commission will probably be needed as a last resort.[28]

RECOMMENDATION 7. FS&T funding should generally favor academic institutions because of their flexibility and inherent quality control, and because they directly link research to education and training in science and engineering.

A distinctive feature underlying the excellence of the U.S. research and development system is the substantial reliance on university-based research (constituting nearly one-third of the FS&T budget for 1994; see Supplement 2). Most of that support is in the form of grants (or grant-like agreements) that support projects initiated by academic researchers and are awarded according to highly competitive merit review. Conducting FS&T at academic institutions has several major benefits:

- It takes advantage of the originality and creativity that students—and their faculty advisors—bring to research;
- It produces exceptionally well prepared scientists and engineers who not only will be the next generation of faculty, but also will work productively in, and transfer technology to, industry and government;
- It allows for easy adjustment of the funding levels in a field because the funding commitment is for a specific project of limited duration;
- It uses merit review to promote the highest quality of work regardless of overall funding levels;
- It draws on academia's own system of reward and recognition, which helps ensure the high quality of the researchers applying for federal grants and keeps them motivated;
- It promotes rapid dissemination of new ideas through the tradition of open publishing and interchange among scholars in academic research (although such interchange is recognized as not being appropriate for classified research);
- It makes research results and expertise widely available to many individuals and private firms, but allows for retention of intellectual property rights to promote commercialization;[29] and
- It builds on well-established and successful collaborations between universities and industry and between universities and federal laboratories.

The committee does not presume that academic research is always of higher quality than that conducted in industry, federal laboratories, or other nonacademic institutions. The committee believes, however, that for most federal science and engineering projects, the distinctive features noted above support a general preference for academic over nonacademic institutions.

Although academic institutions offer many advantages, they can also benefit from a strengthening of their abilities to respond to evolving research opportunities, to maintain emphasis on their educational mission, and to reduce overall costs. For example, the organization of most universities into disciplinary departments can make truly interdisciplinary work difficult to conduct and manage. Projects that require collaboration across units within a university—between organic chemists in a chemistry department and pharmacologists in a medical school, for example—can be more difficult than collaborations among colleagues located at different institutions but working in the same field. Those who pioneer new fields or attempt to bridge research interests among departments or whose work centers on collaboration with other universities, federal laboratories, or industry may risk not being funded or may encounter difficulties in securing space and other resources.

In some research universities and centers, research has overshadowed the educational mission. In response, many universities are placing new emphasis on contributions to education as a criterion in promotion and tenure decisions and are creating interdisciplinary centers that cross traditional departmental boundaries. Indirect costs have been a source of contention between government and universities for many years. Because of budgetary pressures and public concern, however, universities are working with government to reduce costs, including holding down indirect costs and modifying government regulations that can drive them higher.

RECOMMENDATION 8. The federal government should encourage, but not directly fund, private-sector commercial technology development, with two limited exceptions:

- **Development in pursuit of government missions, such as weapons development and spaceflight; or**
- **Development of new enabling, or broadly applicable, technologies for which government is the only funder available.**

The federal government has long sponsored research and education as a means of developing technologies for its own use and has also encouraged the development of state-of-the-art technologies in its capacity as a customer. The histories of the development of airframes and aircraft engines, missiles and satellites, advanced materials, semiconductors, and computers are replete with examples of federal procurement and research support that have contributed to the creation of commercially important technology. Indeed, the government was the first purchaser of key pieces of equipment used to build the components of what has become the Internet.[30] Both FS&T funding and federal procurement will continue to be important in these and other emerging growth sectors linked to federal missions such as health and environmental cleanup. In the future, however, funding for the nation's science and technology base may contribute more to stimulating new sectors of economic growth than will federal procurement and the "demand pull" on an emerging technology.

Even before the end of the Cold War, high-technology spin-offs from federally funded R&D in defense and space had diminished. Efforts have been under way for some time to foster the development of dual-use technologies or to use off-the-shelf commercial technologies in federal programs that develop products for government use. In many cases, civilian applications have now surpassed military ones.

As the Academies' Committee on Science, Engineering, and Public Policy pointed out in its 1993 report, U.S. leadership in high-technology markets cannot be achieved or maintained primarily through federal actions.[31] Commercial technology development will occur largely in the private sector. Firms motivated by market forces and judged by their performance in satisfying demand have a better record than governments of investing in new technologies with large commercial payoffs. As the presumptive owner of the results, the private sector should be the funder of such commercial technology development projects.

The federal government's main role in encouraging commercial technology development and ensuring economic success is to maintain an environment conducive to private-sector development and adoption of new technologies. Such an environment depends on a range of federal policies that influence taxation, macroeconomic stability, national savings, and the volume of international trade. Economic success also is determined by legislation concerned with unfair monopolies, patent protection, product liability, and environmental and consumer protection. Although examination of these critical issues is beyond the scope of this report, the committee believes that government policies, such as those related to taxation, regulation, intellectual property rights protection, social mandates, and others, are usually more important to commercial outcomes than is direct government funding to industry.

The government should not subsidize specific private firms for projects that they would undertake anyway.[32] In a suitable economic context, a firm engaged in product or process innovation will capture or "appropriate" a large fraction of the benefits that it creates. If so, market incentives will guide firms to undertake the right kinds of innovations without any central planning or guidance.

In many cases, however, no one firm can capture the full benefits of its investment. This is generally the case for investment in basic research and can also apply in development related to emerging technologies. One approach to addressing this problem is represented by Sematech, an industry consortium created to improve semiconductor manufacturing, and for which the federal government provided some initial funding. Federal funding may help to establish such consortia in limited and highly specific areas and can be appropriate to support research in consortia formed by industry.

In addition, the government may still have a role in fostering new enabling technologies. Many people believe that nanotechnology (i.e., at scales of one-billionth of a meter) and micromanufacturing, for example, offer exciting commercial opportunities. Government should support training and research that will establish the general scientific and technical principles that firms will ultimately exploit to develop new commercial products and processes. Such investments are appropriate for the federal government because they can generate large benefits that accrue to the nation but would not be captured by any one firm. For example, federal support for research as a component in the education of individuals entering careers in electrical engineering and computer science has helped to produce the skilled people who have developed our modern information technology industries. Support for the work at universities has resulted in the development of the protocols used to exchange information over computer networks, a crucial piece of intellectual capital that all firms have been able to exploit as they enter this new field. Transfer to industry of state-of-the-art technical knowledge produced at science and engineering schools occurs most effectively when faculty, graduate students, and postdoctoral fellows move to the private sector.

Federal funding that improves graduate and undergraduate education is an example of another way to encourage commercial development indirectly, while also supprting R&D in the national interest. In addition to helping stimulate the development and transfer of new enabling technologies into the private sector, the engineering research centers funded by NSF, for instance, have helped change the nature of graduate engineering education.[33] By working in close collaboration with their counterparts in industry, graduate students and faculty have become more aware of the specific technology needs and practices of industry. As a consequence, engineering research programs are more focused and students are better prepared to work in industrial research and development laboratories.

The government also sponsors research and development with potential commercial applications in its own laboratories, in FFRDCs, including the national laboratories, and in independent medical research institutes and other nonprofit organizations (almost half of FS&T funding goes to those organizations, the rest to universities and industrial laboratories). Education is not a central mission of those organizations—an important consideration given that movement of people is one of the most effective ways to transfer new ideas and technologies into the private sector. Several recent reports have noted other reasons that federal laboratories, whether operated by the government or contractors, generally have been less successful than they could be at transferring new enabling technologies to potential users in the private sector.[34] New mechanisms such as cooperative research and development agreements (CRADAs) between firms and the government laboratories were introduced to address this problem. Many successful collaborations have been forged between federal laboratories and industry. Several recent reports argue, however, that CRADAs may be less effective than alternatives, that they are difficult to evaluate because of inadequate data, that ownership of intellectual property is often uncertain, and that they create few jobs.[35,36] Under some CRADAs, the government may be performing research that the partner firm would have done on its own in the absence of a cooperative research agreement. The committee believes that in many cases the government resources that support CRADA research could be better spent on other, more productive items in the FS&T budget.

In addition to providing funds for research and graduate education at universities and government laboratories, the federal government also supports a variety of other programs that promote the development of commercial technologies in the private sector. They include the Advanced Technology Program, the Technology Reinvestment Program, the Manufacturing Extension Partnerships program, Small Business Innovation Research grants and other small business set-asides, and direct government subsidy to private firms. Those programs have different goals and structures but share in their intention to cultivate industrial innovation. The ATP and the TRP involve funding of private-sector projects; the MEP program is modeled after the agricultural extension service program and primarily helps small businesses to incorporate new technologies (see Supplement 1). Most of these programs are too new to be carefully evaluated, and, because of inherent features in program design and prospects of unstable funding, we may never be able to tell whether some of them achieved their goals.[37]

At this time, the very concept of a government role in subsidizing the development of private-sector product and process development is controversial. Some difficult questions arise with subsidized partnership programs such as the ATP—will they succeed in fostering new, commercially relevant technologies that otherwise would not develop as quickly, and are they the most efficient uses of increasingly scarce federal R&D dollars? The committee is skeptical that the answer to these questions is yes. It therefore believes that these subsidized industrial partnership programs should be continued only if the case is convincingly made that the government is the funder of last resort for an important enabling technology, and they should be pursued only on an experimental basis, with careful attention to their goals, the distribution of proprietary rights, and how they will be evaluated. Where a new technology is needed to address a specific mission such as a military need, however, federal leadership is better justified, as noted in the first bulleted item under Recommendation 8.

RECOMMENDATION 9. **FS&T budget decisions should give preference to funding projects and people rather than institutions. That approach will increase the flexibility in responding to new opportunities and changing conditions.**

Compared to most other developed countries, the United States awards a higher fraction of its research and development funding to specific projects as opposed to distributing funds through institutional or formula grants. This mode of funding has several important advantages. It promotes the scientific and technical quality and originality of proposals; it permits awards to be made on the basis of competitive merit review procedures; and, by investing in projects and people rather than institutions, it makes the research and development system more flexible and responsive to changing scientific opportunities and national needs. Together those features have created a broad base of first-rank research institutions across the country that have adapted to major shifts in federal research and development priorities over time.

The committee strongly endorses the principle of favoring the support of projects and people over institutions. The pace of scientific discovery has quickened and the time from discovery to innovation and commercialization is becoming shorter in many fields, which makes the flexibility and responsiveness of the research and development system increasingly crucial. To free up or reallocate resources to meet new opportunities and needs, it is much easier to cut back or eliminate a program of project grants than it is to disengage from support of institutions. If an agency's budget is cut, there is a danger that funds will be taken automatically from its extramural program. Instead, the available funds should be allocated to those people and projects best able to accomplish the task—whether in universities, federal laboratories, or other institutions.

In the future, there should be a presumption against establishing new permanent institutions. Moreover, the establishment of any such institutions and major programs or centers should include a time limit or "sunset" provision, along with periodic review.

Within the General Constraints Determined by National Priorities, the Selection of Individual Projects Must Reflect the Standards of the Scientific and Technical Community.
(Recommendations 10 and 11)

RECOMMENDATION 10. Because competition for funding is vital to maintain the high quality of FS&T programs, competitive merit review, especially that involving external reviewers, should be the preferred way to make awards.

The highest-quality projects and people should be supported with FS&T funds. The best-known mechanism to accomplish that is some form of open competition involving evaluation of merit by peers. Competitive merit review involves the use of criteria that include technical quality, the qualifications of the proposer, relevance and educational impacts of the proposed project, and other factors pertaining to research goals rather than to political or other nonresearch considerations.[38] Open competition means that, at some level within the framework of an agency's mission, researchers propose their best ideas and anyone may apply and be funded regardless of institution or geographic location. However, in the case of highly targeted missions, quality can also be maintained by knowledgeable program managers who have established external scientific and technical advisory groups to help assess quality and to help monitor whether agency needs are met (see Supplement 3 and Box II.8).

The committee believes that the principle of merit review—which emphasizes competition among ideas, diversity of funders and performers of research and development, and organizational flexibility—has been largely responsible for the remarkable quality, productivity, and originality of U.S. science and technology in the past. Competitive merit review should be the method of choice for making future decisions about FS&T funding.

Many federal research and development agencies have developed some form of competitive merit review process to use in making extramural awards for research, training, and facilities. They have also worked to develop equivalent systems of review for allocating intramural funding, but merit review of in-house research is much more difficult because federal research scientists and engineers are in the civil service and still retain salary and benefits even if they are not productive or their area has lower priority or has become obsolete. That problem is a perennial one in the periodic reviews of federal laboratories.[39] The FFRDCs, including the national laboratories, also have procedures for allocating research funding competitively based on performance. Some do it well, but overall the results have been uneven.[40]

There are other approaches to promoting high quality in federally supported research and development. Some programs try to identify top researchers and give them long-term support rather than require them to submit specific proposals to compete every few years. Some funding for agricultural research is allocated to state agricultural experiment stations and land-grant colleges on a formula basis, and the supported institutions choose the researchers and their projects. Evaluations of that system of formula-grant allocation have not given high marks to its responsiveness or the quality of the resulting research.[41] Other federal funding is awarded competitively to research centers, which in turn distribute the funding among individual researchers and groups.

There is benefit to having a variety of approaches to supporting FS&T, especially because mission agencies have specialized assignments to fulfill. However, the committee believes that fiscal constraint makes it important to level the playing field. Competitive merit review should therefore be increased relative to other mechanisms for awarding FS&T funds. Merit review is best exemplified by the processes used at the NSF and NIH, that is, the use of external peer review to identify and select the best proposals for individual research projects as part of a review process based on competition and expert evaluation of merit criteria. That approach enables those two agencies to choose the best performers. Accordingly, use of competitive merit review to allocate federal funding should be the default presumption, supplemented with other mechanisms for inherently governmental functions that cannot be accomplished through competitive merit review.

RECOMMENDATION 11. Evaluations of research and development programs and of those performing and sponsoring the work also should incorporate the views of outside evaluators.

Technical merit, which is the primary criterion used in performance reviews of research agencies and programs as well as proposals, is best evaluated by independent scientific or engineering peers. Agency performance review systems differ in the extent to which they use external reviewers, but there are two compelling reasons to rely heavily (although not exclusively) on external reviews. First, because the federal government funds most research and development outside its own laboratories in industry, universities, and other nongovernment research institutions, most of the qualified reviewers are outside government. Second, external reviewers are a more diversified source of opinion and can bring a wider range of experiences to the review process compared with federal agency personnel. Where needs are highly specific, such as development of a stealth aircraft or rapid response to an emerging infection, external reviews are still useful, although they may have to be retrospective rather than prospective. Government officials must make the final decision.

Recent changes across the federal government emphasize improving performance review and program evaluation. Indeed, according to the Government Performance and Results Act (GPRA) of 1993 (Public Law 103-62), every federal agency must have performance goals and measures for its programs (including FS&T programs) by 1997 for its Fiscal Year 1999 budget submission. It will be difficult to apply GPRA requirements to research and development activities because, by their nature, they are long-term and their impacts are diffuse and hard to measure.[42]

Any system to allocate resources should be guided by explicit goals, expressing the underlying philosophy and criteria for evaluating performance. But a clear message emerges from the abundant recent writing on applying performance measures to research and development: it is a complicated business. The science of metrics documents that most measures are incomplete, and mindless application actually can undermine the very functions such measures are intended to improve.[43] Just as the tyranny of quarterly bottom lines can frustrate long-term corporate planning, so also can science be distorted by simple indicators such as publication counts, citation counts, patent counts, doctorates produced, or user satisfaction ratings. These are useful, but incomplete, measures. Several recent assessments of such measures concluded that they must be augmented by expert judgment.[44] One review observed that such measures may leave out "virtually all of what researchers themselves find important about their work. One could have a government full of programs that performed beautifully according to these indicators, and still be at the trailing edge of every scientific frontier."[45]

It makes sense to track relevant measures, but they cannot supplant the essential element of expert judgment that is the bedrock of quality assessment in research and development. Scientists and engineers seeking federal support should be accountable to the public, and the standards should capture what constitutes the best science and engineering. To the extent that performance review and program evaluation come into wider use in assessing FS&T funded activities, they will have to incorporate expert judgment of quality, impact, and other important aspects that will benefit from the use of outside reviewers.[46]

Ideally, in government as in the private sector, every organization should ask basic questions about the need for its continued existence on a regular basis. In one formulation, every department and agency and each subunit and activity should answer the following questions satisfactorily:[47] What is our mission? Is it still the right mission? Is it still worth doing? If we were not already pursuing this mission, would we still choose it now?

In most cases, agencies are responding to statutes, congressional report language, or presidential initiatives. These questions, therefore, may need to be raised at more than just the agency level.

The Federal Government Must Implement a Structure Capable of Fostering, Not Hindering, the Management of Research and Development. (Recommendations 12 and 13)

RECOMMENDATION 12. Research and development should be well managed and accountable but should not be micromanaged or hobbled by rules and regulations that have little social benefit.

Science and technology must be managed well, particularly when public funds are at stake. Fraud, misuse of funds, violations of human subject protections, or other abuses should not be tolerated. Maintaining safeguards requires credible mechanisms for investigation and enforcement. At the same time, federal agencies must strike a balance between the need for accountability and the burden of regulation. Public dissemination of the results of federally funded research and development is an important element in achieving maximum return on public investment, and it also contributes to defining for the public the value of that investment.

If there is no regulation, the risk of abuse will rise, but regulation imposes significant cost. In the past 2 decades, the trend has been toward increased paperwork to comply with procurement regulations, fair hiring practices, restrictions on drug use, and many other public concerns that are important but that impose constraints on the conduct of federally funded research and development.[48]

Because procedures intended to enhance accountability have become increasingly burdensome, continued scrutiny of the purposes, effectiveness, costs, and alternatives to current practices would be welcome, beginning with a thorough overhaul of the regulations and followed by systematic, periodic reviews. The Office of Management and Budget and the Office of Science and Technology Policy should work together to target one or a few areas of regulation and accountability assessment each year and should encourage agency innovation to streamline or replace current practices.

The effect of regulations and social mandates can be quite severe for performers of federally funded research and development. If regulations are reviewed and either reduced, streamlined, or eliminated by the OMB-OSTP effort recommended, the committee believes that the productivity of the research and development system can be improved and costs can be reduced. For their part, universities and other performers should review their own procedures and regulations. The Federal Demonstration Project sponsored by the Academies' Government-University-Industry Research Roundtable demonstrates that improvements can be made without sacrificing important goals.[49]

RECOMMENDATION 13. The federal government should retain the capacity to perform research and development within agencies whose missions require it. The nation should maintain its resulting flexible and pluralistic system of support. The executive and legislative branches should implement the procedures outlined in the committee's Recommendations 1 through 4 to ensure a more coherent FS&T budget process whether or not a Department of Science is established.

Any changes in the structure of federal support for science and technology should take into account the linkage between research and development and agency missions and the benefits derived from a robust and pluralistic R&D system. Most federally funded research and development is conducted in pursuit of national goals such as a strong defense, better health, exploration of space, wiser use of natural resources, and greater agricultural production (see Supplements 1 and 2). This linkage to government agency missions is a strength of the U.S. research enterprise and has produced a robust and pluralistic R&D support system. Other than basic research programs at the National Science Foundation, few federal science and technology programs have been set up to support research as an end in itself. Even the National Science Foundation has an educational mission in addition to its support of science and engineering. Given their purpose, agency programs are and should be evaluated first for their contribution to their departments' goals and only later for their place in a balanced national research and development system.[50]

Current proposals for a Department of Science in part follow this principle by leaving most militarily relevant research and development in the Department of Defense, health research in the Department of Health and Human Services, and agricultural research and development in the Department of Agriculture. While wisely retaining research and development in mission agencies, this approach would limit a Department of Science to activities that fall outside existing mission agencies. Such a Department of Science would have a smaller research budget than the National Institutes of Health and a significantly smaller development budget than the Department of Defense.

Creating a Department of Science because cabinet departments are abolished or reconfigured, rather than as a result of applying criteria for allocating federal funds for research and development, involves considerations beyond the charge to this committee. Such a Department of Science, however, cannot fully address the need for review, coordination, and FS&T budget allocation among departments. The committee believes that its recommendations will contribute more to planning, coordinating, and evaluating federal science and technology than either the current system or a Department of Science.

The growth of federal science and technology from multiple roots in mission agencies has resulted in a pluralistic research and development system. Although some may see needless overlap in such a system, in reality pluralism is a great source of strength, an advantage over the ways research and development are organized in many other countries. The diversity of performers fosters creativity and innovation. It increases the number of perspectives on a problem. It makes competition among proposals richer, and it induces competition to support the best work among funders, both public and private. At the same time, diverse funding alternatives give original ideas a better chance to find support than would a more centralized system. A pluralistic research and development system thus enhances quality and our national capacity to respond to new opportunities and changing national needs. The challenge in the current period is to retain diversity and balance while cutting back in some areas to free resources for better or more important activities.

As emphasized in Recommendation 1, integrating the needs of a pluralistic research and development system across multiple agencies and programs requires a comprehensive overview and careful planning. The federal budget process should take into account how interdependent different fields of science and technology have in fact become. The impact of cutbacks in one agency on major fields, on other agencies, and on national goals should be considered. Changing or scaling back an agency's mission (e.g., to reduce and reorient the post-Cold War defense establishment) generally has implications for the type and scale of research and development it, and others, conduct. As noted above, for example, DOD provides most of the federal funding for academic research in several engineering fields and computer science. Computer-intensive biological research supported by NIH and NSF, such as genome research or structural analysis for drug design, could thus be affected by cuts in DOD computer science. Important advances and efficiencies enabled by increasingly powerful computation and by use of the Internet and global communications supported by many agencies could also be impeded by such cuts.

Monitoring the impact of cuts in one part of the research and development system on another part is a function that the current budget process does not perform systematically. Cross-program impacts are accommodated to some extent in the decentralized negotiations of budget line items in individual agencies, and special initiatives often identify items in multiple agencies. Cross-agency planning is not routine, however, even in the limited sense of "damage control" that is important when budget cuts are contemplated, and the FS&T budget is not monitored as a whole as the budget process unfolds. The committee's Recommendations 1 through 3 in effect give the President's Science and Technology Advisor and the Office of Management and Budget a strong integrative role, with the authority to effect transfers across departments and agencies that no cabinet official can perform. The recommendations also entail monitoring the FS&T budget as a whole in Congress, beyond that fraction that might be included in a Department of Science. If the recommended process is used in tandem with the principle of retaining world leadership embodied in Recommendation 4, the federal government will have a more coherent and effective research and development system.

Looking to the Future

A robust national system of innovation lies at the heart of our economy, our health, and our national security. That system of innovation depends on federal investments. The committee believes that its recommendations address a crucial need: maintaining the strength and vigor of U.S. research and development despite the prospect of declining federal discretionary spending over the next several years. Seeing the science and technology enterprise through the lens of a unified FS&T budget can help leaders in government and the American public to gauge its fiscal health. A carefully constructed comprehensive budget offers a unitary view, not artificially balkanized into agency budgets, but sensitive to the complexities and relationships among government programs vital to maintaining the United States at the forefront of world-class science and technology. The corollary proposals provide the basis for continuing excellence—emphasizing programs and people rather than institutions, subjecting all federal science and technology activities to competitive merit review, linking science and engineering research to education, and maintaining a pluralistic system of research and development tied to public missions. The committee's recommendations are designed to help root out obsolete or noncompetitive activities, allowing good programs to be replaced by even better ones.

Science and technology have utterly transformed our world over the past 50 years, touching almost every aspect of our daily lives—from communication to transportation to health (Box I.5). They will be at least as important over the next half century. Preeminence in science and technology has become a national asset, at once a point of pride and an immensely practical investment. Prudent stewardship of science and technology, as much as any other area of federal policy, will dictate how our children and our grandchildren live.

Box I.5
Lessons from the Past and Some Opportunities for the Future

Though enormously visionary, the scientists and political leaders who set the United States on its post-World War II research and development course could never have foreseen the extraordinary results. The computer was in its infancy in 1945 and seemed more a research tool than a revolutionary device that would profoundly affect industry, commerce, the financial world, government, science, education, communications, entertainment, and society as a whole. Accurate weather forecasting covered about a day in 1945; reliable 3- and 6-day forecasts, and the 90-day outlooks now relied on by farmers and utility companies, came only with years of research and the advent of supercomputers. Microelectronics, with all its implications for space exploration and utilization, national security, consumer electronics, medicine, and domestic and international communications, did not exist—nor did the equally revolutionary laser. Materials science, given a boost by the war, had yet to benefit from the studies that would yield the new metal alloys, high-strength steels, composite materials, silicon chips, glassy metals, optical fibers, and polymers so vital and so valued in 1995.

Astronomy meant mostly optical telescopes at war's end, and astronomers could only dream of the striking images now provided daily by the Hubble Space Telescope; the great advances provided by radio, infrared, ultraviolet, X-ray, and gamma-ray astronomy would come only with time. Though an early cosmological vision of the universe's birth existed, it had yet to win its popular name, "The Big Bang," or to gain the theoretical underpinnings and experimental backing that now make it the standard model for the cosmos's origin. The Earth's crust was accepted as a solid shell, not the giant, separate blocks of rock portrayed by the theory of plate tectonics, which came together in the 1950s and 1960s and provided earth scientists with a general framework to explain the cause of most giant earthquakes, why volcanoes exist where they do, the birth of new oceans, and the timeless drifting of the continents around the globe. Few paid attention to or realized the economic, health, and social implications of a deteriorating environment, the loss of biodiversity, or the potential for adverse climate change—vital world issues that researchers would identify, describe, and bring to public attention.

The personal computer first appeared in the 1970s; the explosive growth of the Internet is a 1990s phenomenon. Electronic mail was until very recently the tool of a narrow slice of the scientific and technical community. Now, our national security depends heavily on the use of computers, networks, and telecommunications to assess, understand, and respond to potential threats. Computer graphics provides the "vision" to design new materials and buildings, and to model, for example, the lethal process of an AIDS (HIV) virus entering a cell and co-opting its functions. There is virtually no industry that is not being transformed by the information revolution. And yet, the information revolution is still young and hardly over.

The remarkable advances enabled by science and technology during the past 50 years will surely be extended in the next 50. We can see some of the outlines. Information technology, for example, is already transforming the operations of many of our basic institutions, offering new ways to educate our children and contributing new approaches and tools for research in science and technology. Less obvious is how a quickly widening range of challenges facing our nation and the world will be addressed. If history is a guide, the work now under way in universities and in federal and industrial laboratories will play a vital role.

The health challenges to the nation are apparent. The population is aging, and with that the problems of heart disease, cancer, and degenerative illnesses such as Alzheimer's disease appear in sharp relief. These illnesses require fundamental understanding not only of the underlying biology but also of effective prevention strategies to delay or block their onset. The problem of "emergent diseases" has gained full force in this decade, from the resurgence of tuberculosis to the appearance of "jet-age" scourges, such as AIDS and Ebola virus. We can rightly take comfort in the past victories over polio and smallpox and other infectious diseases. We should not forget, however, that the polio vaccine built on a century of microbiology, that

continued on next page

biotechnology is only now becoming central to drug discovery, and that the biology underlying many of today's dread diseases is still almost wholly unknown. Further, science and technology are essential to building on the effective campaigns to reduce infant mortality, smoking, and deaths and injuries from drunk driving.

Perhaps less obvious but just as promising is the future potential for science and technology to address diverse national needs in transportation, public infrastructure, agriculture, and the environment. New materials, propulsion systems, and imaginative use of information technologies to create "smart" highways and cars will map onto currently obvious transportation needs—from reducing pollution to improving traffic flow and highway design. Research has contributed, albeit considerably below its potential, to development of the national systems by which we get our drinking water, remove our wastes, and obtain electrical power. As these systems become more complex and the pressures on public funds intensify, research that reduces costs and improves safety, such as non-destructive testing of bridges, tunnels, railroad tracks, and the like, will become even more urgent.

U.S. agriculture has been a triumph. Now the advent of biotechnology has created major new opportunities to increase the quality of foods, raise the efficiency of crop production, and develop new industrial uses for crops, including biodegradable plastics and pharmaceutical products. The current U.S. export lead in agriculture builds on a century of public and private investments in agricultural research and development. Future research will surely offer ways to sustain the productivity of U.S. agriculture while also making it more environmentally benign.

Finally, resource pressures will inexorably increase as we enter the next millennium—as populations, industrialization, and demand for energy and other resources increase. These pressures will increase debates about risks versus costs. Informing that debate will require a base of science and technology so that the problems are well understood, the impacts of alternative remediation strategies are analyzed, risks are adequately assessed, and effective prevention strategies are put into place.

A strong research and development capacity will be integral to dealing with future challenges, whether environmental problems, medical emergencies, or national security threats—or crises that we cannot yet predict. We also know that solutions come in unexpected ways from what is the world's premier research enterprise. With wise management, solutions to pressing problems—and innovations giving rise to now unimagined advances—will continue to come from many directions, for example, from the work of astronomers trying to understand the large-scale structure of the universe, or from mathematicians' studies on improving the calculations of properties of alloys, or from the efforts of social scientists to devise new ways for institutions to manage public resources such as fisheries, grazing grounds, and water supplies, or from biologists' investigations of the neural systems of invertebrates. New knowledge that enlarges our understanding will in time serve national needs. Science and technology, contributing a unique national capability for problem solving and creative discovery, will continue to be key in keeping the United States in its world leadership position—economically, militarily, and intellectually.

Endnotes

1. *The Budget of the United States Government, Fiscal Year 1996,* Chapter 7, "Investing in Science and Technology" (Washington, D.C.: U.S. Government Printing Office, 1995), p. 94.

2. The phrases *research and development* and *science and technology* are often used interchangeably. The committee has chosen to use *research and development*, except when it is explicitly referring to its proposed budget index, federal science and technology (FS&T), and the work encompassed by it.

3. The interdepartmental coordination mechanism was the Federal Coordinating Council for Science and Technology under Presidents Reagan and Bush, and now is the National Science and Technology Council under President Clinton.

4. Calculated from Tables C-61 and C-62 in National Science Foundation, *Federal Funds for Research and Development, Fiscal Years 1993, 1994, and 1995*, NSF 95-334 (Arlington, Va.: NSF/Division of Science Resources Studies, forthcoming). In 1994, the Department of Defense funded 59 percent of academic research in electrical engineering, 69 percent in metallurgy and materials science, and 56 percent in computer science.

5. The Bush and Clinton administrations initiated structures and procedures that begin to implement several of the processes and criteria listed. The last budgets of the Bush administration included agency "cross-cuts" that took into account multiagency initiatives. The Clinton administration continued this practice and also prepared a separate chapter on research and development as part of the President's budget (*The Budget of the United States Government, Fiscal Year 1996*, Chapter 7, "Investing in Science and Technology," 1995). The many activities of the National Science and Technology Council are summarized in its "Accomplishments Report, 1993-1995," Office of Science and Technology Policy, Executive Office of the President, 1995.

6. Allen Schick, *The Federal Budget: Politics, Policy, Process* (Washington, D.C.: The Brookings Institution, 1995); Willis H. Shapley, *The Budget Process and R&D* (New York: Carnegie Commission on Science, Technology, and Government, 1992).

7. Carnegie Commission on Science, Technology, and Government, *Science, Technology, and Congress: Organization and Procedural Reforms* (New York: Carnegie Commission on Science, Technology, and Government, 1994).

8. These criteria are adapted from the Committee on Science, Engineering, and Public Policy (National Academy of Sciences, National Academy of Engineering, and Institute of Medicine), *Science, Technology and the Federal Government: National Goals for a New Era* (Washington, D.C.: National Academy Press, 1993).

9. Throughout this report, the term *federal laboratories* refers to laboratories owned and operated by the federal government (including intramural laboratories), laboratories owned by the federal government but operated by contractors (including the national laboratories administered by DOE), and other FFRDCs. See Box II.6 for an explanation.

10. Department of Defense, *Department of Defense Response to NSTC/PRD 1, Presidential Review Directive on an Interagency Review of Federal Laboratories*, February 24, 1995; Department of Defense, *Draft Interim Report to the National Science and Technology Council, Presidential Review Directive 1*, October 12, 1994; Defense Science Board, *Laboratory Management Interim Report*, background for the Base Closure and Realignment 1995 (BRAC 95 Addendum), April 3, 1995. Collectively, these three reports are known as the Dorman Report.
NASA Federal Laboratory Review Task Force, NASA Advisory Council, *NASA Federal Laboratory Review* (Foster Report) (Washington, D.C.: NASA, February 1995).
Task Force on Alternative Futures for the DOE National Laboratories, *Alternative Futures for the Department of Energy National Laboratories* (Galvin Report) (Washington, D.C.: Department of Energy, February 1995).
Ad Hoc Working Group of the National Cancer Advisory Board, *A Review of the Intramural Program of the National Cancer Institute* (Bishop/Calabresi Report) (Bethesda, Md.: National Institutes of Health, June 26, 1995); External Advisory Committee of the Director's Advisory Committee, *The Intramural Research Program* (Cassell/Marks Report) (Bethesda, Md.: National Institutes of Health, April 11, 1994).

National Science and Technology Council, *Interagency Federal Laboratory Review, Final Report* (NSTC Report) (Washington, D.C.: Office of Science and Technology Policy, May 15, 1995).

11. Federal Laboratory Review Panel, *Report of the White House Science Council* (Packard Report) (Washington, D.C.: Office of Science and Technology Policy, May 1983); Bishop/Calabresi Report, 1995; Dorman Report, 1995.

12. Alan K. Campbell, Stephen J. Lukasik, and Michael G.H. McGeary, eds., *Improving the Recruitment, Retention, and Utilization of Federal Scientists and Engineers* (Washington, D.C.: National Academy Press, 1993); Foster Report, 1995; Dorman Report, 1995; Cassell/Marks Report, 1994; Alan L. Dean and Harold Seidman, *Options for Organizational and Management Reform for the Intramural Research Program of the National Institutes of Health* (Washington, D.C.: National Academy of Public Administration, July 1988); Institute of Medicine, *A Healthy NIH Intramural Program: Structural Change or Administrative Remedies?* (Washington, D.C.: National Academy Press, 1988).

13. Foster Report, 1995, p. 9.

14. Michael E. Davey, *DOD's Federally Funded Research and Development Centers,* CRS Report for Congress 95-489, Science Policy Research Division, Library of Congress (Washington, D.C.: Congressional Research Service, April 13, 1995); Office of Technology Assessment, *Department of Defense Federally Funded Research and Development Centers* (Washington, D.C.: U.S. Government Printing Office, June 1995); Defense Science Board Task Force, *The Role of Federally Funded Research & Development Centers in the Mission of the Department of Defense* (Washington, D.C.: Office of the Under Secretary of Defense for Acquisition and Technology, April 1995).

15. Dorman Report, 1995; Foster Report, 1995; Galvin Report, 1995; Packard Report, 1983; NSTC Report, 1995.

16. Calculated from Table C-154a in National Science Foundation, *Federal Funds for Research and Development: Fiscal Years 1992, 1993, and 1994,* NSF 94-328 (Arlington, Va.: NSF/Division of Science Resources Studies, 1995).

17. Defense Science Board Task Force, *The Role of Federally Funded Research & Development Centers in the Mission of the Department of Defense,* 1995.

18. The Foster Report (1995) specifically recommends that NASA laboratories reduce their insularity, enhance ties with universities, and adopt the personnel and management practices of the only major NASA FFRDC, the Jet Propulsion Laboratory, which is associated with the California Institute of Technology.

19. Galvin Report, 1995, p. 4.

20. Galvin Report, 1995. Other analyses of the DOE-supported national laboratories and their futures have concluded that carefully planned diversification could be useful if done well. Barry Bozeman (Georgia Institute of Technology) and Michael Crow (Columbia University), who drew on a large body of past research in a report for the Department of Commerce on the role of all federal laboratories (*Federal Laboratories in the National Innovation System: Policy Implications of the National Comparative Research and Development Project,* May 1995), note that the purposes of various laboratories vary tremendously. Ann Markusen and colleagues at Rutgers University examined Sandia and Los Alamos National Laboratories and concluded that there is a restricted stock of knowledge and technology that is no longer secret (Ann Markusen, James Raffel, Michael Oden, and Marlen Llanes, *Coming in from the Cold: The Future of Los Alamos and Sandia National Labora-*

tories, Piscataway, N.J.: Center for Urban Policy Research, 1995). Both the Bozeman/Crow and Markusen et al. reports are more open to consideration of new mission areas than is the Galvin Report, although Markusen et al. specifically note that the two DOE laboratories in New Mexico should be reduced in size. Markusen is quite critical of the Galvin task force recommendation that laboratories engaged in weapons design and nuclear cleanup activities be delegated to autonomous "corporatized" units. However, the Bozeman/Crow and Markusen reports both support the main thrust of the Galvin task force—that current methods of technology transfer are poorly understood and probably inefficient—and concur with the notion that national laboratories should compete with universities and private performers, rather than have unique access to funds, and should strengthen their ties to one another, to academia, and to industry.

21. Galvin Report, 1995, pp. 8-10, 55.

22. Federally funded research and development centers have a long-term contractual relationship with the government and are operated by contractors—see Boxes II.5 and II.6, Supplement 2. See also U.S. Congress, Office of Technology Assessment, *A History of the Department of Defense Federally Funded Research and Development Centers* (Washington, D.C.: Government Printing Office, July 1995 [GPO Stock No. 052-003-01420-3]).

23. National Science and Technology Council, *Interagency Federal Laboratory Review, Final Report* (NSTC Report), 1995. See especially pp. 9-19 and 21-22.

24. NSTC Report, 1995.

25. Dorman Report, 1995.

26. Foster Report, 1995; Galvin Report, 1995. Also, NSTC Report, 1995.

27. Bishop/Calabresi Report, 1995.

28. In 1990, Congress passed and the President signed the Defense Base Closure and Realignment Act, which was intended to protect the base-closing process from electoral politics and to insulate the President and members of Congress from difficult decisions that affect important political constituencies, particularly local groups adversely affected by closures and shrinkage. The act established an independent commission to hold public hearings, take recommendations from the Secretary of Defense, and make a list of recommended closures and reconfigurations. Under terms of the act, the President then may accept or reject the entire list but cannot add or delete specific items on it, and Congress has 45 days to veto his action. In the absence of action by the President or Congress, the commission's list becomes the basis for closures and realignments by the Department of Defense. A total of 250 installations were closed or reduced through three rounds of closings—in 1991 and 1993, plus another round conducted in 1988 under a previous law. The final round, which included some laboratory facilities, was completed in 1995.

The currently established BRAC process is conducted by a single cabinet-level department, and the commission's sole responsibility is to make recommendations about closing and shrinking facilities, not about reallocation across departments. Depending on its purpose and scope, however, a laboratory closing commission might cut across departments and independent agencies, and might be asked to reallocate as well as close or reduce facilities, complicating its task and opening the question of the proper venue for organizing such actions.

29. Pursuant to the Bayh-Dole Act of 1980 and subsequent amendments and executive orders, academic centers retain patent rights and copyrights that result from federal funding, with certain restrictions. Those rights can be licensed to one or more firms, depending on the nature of the inventions or other results. Moreover, because they are held by universities, the licensing arrange-

ments are subject to greater public scrutiny than is direct federal funding to private firms with patents held by those firms.

30. Computer Science and Telecommunications Board, National Research Council, *Realizing the Information Future: The Internet and Beyond* (Washington, D.C.: National Academy Press, 1994).

31. Committee on Science, Engineering, and Public Policy (National Academy of Sciences, National Academy of Engineering, and Institute of Medicine), "The Federal Role in the Development and Adoption of Technology," Chapter 4 in *Science, Technology and the Federal Government: National Goals for a New Era*, 1993, pp. 31-44.

32. That principle is perhaps best exemplified in defense technologies developed with federal funding, which can now be more expensive and less advanced than commercial technologies. Recent attention to acquiring dual-use technologies from commercial sources and exploiting defense technologies in commercial markets stems from this reversal in the traditional flow of new technology. This circumstance is noted by the Committee for National Security of the National Science and Technology Council, in *National Security Science and Technology Strategy* (Washington, D.C.: Office of Science and Technology Policy, 1995). See also National Economic Council, National Security Council, and Office of Science and Technology Policy, Executive Office of the President, *Second to None: Preserving America's Military Advantage Through Dual-Use Technology*, Doc. No. ADA 286-779 (Fort Belvoir, Va.: Defense Technical Information Center, February 1995).

33. Engineering Centers Division, Directorate for Engineering, National Science Foundation, *The ERCs: A Partnership for Competitiveness*, NSF 991-9, 1991; *Highlights of Engineering Research Centers Technology Transfer*, NSF 92-6, 1992; and *Highlights of Engineering Research Centers Education Programs*, NSF 95-56, 1995 (Arlington, Va.: National Science Foundation).

34. The Foster Report (1995) at several points notes that NASA laboratories are "insular" and that in many areas private firms have raced ahead of parallel NASA programs. Other reports also cited above, most notably the Galvin Report (1995) and those by Bozeman and Crow and by Markusen et al. (note 20), also point to difficulties in technology transfer from federal laboratories.

35. Start-up of new firms, for example, has been a major source of innovation. The importance of a "culture" that nurtures innovation has been stressed in several recent works, including that of Ann Markusen et al. (note 20); Susan Rosegrant and David R. Lampe, *Route 128: Lessons from Boston's High-Tech Community* (New York: Basic Books, 1992); AnnaLee Saxenian, *Regional Advantage: Culture and Competition in Silicon Valley and Route 128* (Cambridge, Mass.: Harvard University Press, 1994); and Economics Department, Bank of Boston, *MIT: Growing Businesses for the Future* (Boston: Bank of Boston, 1989). Markusen et al., in particular, suggest that policies to encourage movement of experts and technologies out of government laboratories might well be more effective than those, such as CRADAs, that retain talent and technology within laboratory walls.

36. The Galvin Report (1995) concludes that CRADAs may have distracted DOE's major multipurpose national laboratories from their central missions. The report by Markusen et al. (note 20) suggests that CRADAs may be less effective than less expensive and more historically important means of technology transfer, and points out that evaluation is hampered by poor access to data. Bozeman and Crow (note 20) observe that 90 percent of CRADAs produce no jobs, and also note that a "one-size-fits-all" technology transfer policy for federal laboratories flies in the face of their diversity. The place of CRADAs in DOE national laboratories is an area of active controversy (see Colin Macilwain, "US Weapons Labs Face Curb on Civilian Role," *Nature* 376 (July 13): 106-107, 1995).

37. Evaluation of investment programs to date has focused mainly on the question, Would this technology ever have developed or would it have been significantly delayed but for the federal funding? Most assessments have been based on queries to recipients and agency staff about judgments of success, and on limited measures of impact such as patent counts or financial measures that cannot answer the question. What is needed is rigorous assessment through comparison to appropriate control cases. Moreover, answering one question does not address several others that are equally important, such as: How effective is direct federal investment in specific firms or consortia compared to investment in R&D through other mechanisms, such as grants and contracts to do similar work at universities or federal laboratories? Would incentives to R&D performers to ease start-up of new firms or to encourage private investment through indirect means achieve the same ends at less cost or with less direct federal involvement? How can direct investments in firms or consortia confer proprietary advantage and yet ensure public accountability and fair access by other firms to data, results, and expertise?

38. NSF Advisory Committee on Merit Review, *Final Report*, NSF 86-93 (Washington, D.C.: National Science Foundation, 1986).

39. Dorman Report, 1995; Galvin Report, 1995; Foster Report, 1995; Bishop/Calabresi Report, 1995; and Cassell/Marks Report, 1994.

40. Reviews of federal laboratories consistently conclude that procedures for judging the quality of research are not adequate and in practice do not have much effect on the allocation of research funding. See, for example, Defense Science Board, *Laboratory Management Interim Report*, 1995 (note 10); Bishop/Calabresi Report, 1995; Foster Report, 1995; Cassell/Marks Report, 1994; National Research Council, *Interim Report of the Committee on Research and Peer Review in EPA* (Washington, D.C.: National Academy Press, 1995); Carnegie Commission on Science, Technology, and Government, *Environmental Research and Development: Strengthening the Federal Infrastructure* (New York: Carnegie Commission on Science, Technology, and Government, 1992); U.S. Environmental Protection Agency, *Safeguarding the Future: Credible Science, Credible Decisions*, EPA/600/9-91/050, Expert Panel on the Role of Science at EPA (Washington, D.C.: U.S. Government Printing Office, 1994).

41. See, for example, National Research Council, *Investing in the National Research Initiative: An Update of the Competitive Grants Program in the U.S. Department of Agriculture* (Washington, D.C.: National Academy Press, 1994).

42. Susan E. Cozzens, "Assessment of Fundamental Science Programs in the Context of the Government Performance and Results Act (GPRA)," RAND Project Memorandum PM-417-OSTP (Washington, D.C.: Critical Technologies Institute, 1995).

43. Thomas D. Cook and William R. Shadish, "Program Evaluation: The Worldly Science," *Annual Reviews of Psychology* 37: 193-232, 1986; Robert K. Merton, *The Sociology of Science: Theoretical and Empirical Investigations* (Chicago: University of Chicago Press, 1973); P.H. Rossi, H.E. Freeman, and S. Rosenbaum, *Evaluation: A Systematic Approach* (Beverly Hills, Calif.: Sage Publications, 1982).

44. Cozzens, "Assessment of Fundamental Science Programs," 1995; Commission on Physical Sciences, Mathematics, and Applications of the National Research Council, *Quantitative Assessments of the Physical and Mathematical Sciences: A Summary of Lessons Learned* (Washington, D.C.: National Academy Press, 1994); Susan E. Cozzens, rapporteur, *Evaluation of Fundamental Research Programs: A Review of the Issues*, discussion draft, Office of Science and Technology Policy, August 15, 1994; and Susan Cozzens, Steven Popper, James Bonomo, Kei Koizumi, and Ann Flanagan,

Methods for Evaluating Fundamental Science, DRU-875/2-CTI, Critical Technologies Institute, RAND Corp., for the Office of Science and Technology Policy, October 1994.

45. Cozzens, "Assessment of Fundamental Science Programs," 1995, p. 33.

46. Research programs should be evaluated at a fairly aggregate level by independent individuals with the requisite scientific and technical expertise, who are capable of judging progress relative to resources invested. By "a fairly aggregate level," the committee means including a fairly large set of projects and over a sufficient period to capture benefits, which are often long delayed; the more basic the science, the longer the gestation period. Scientists must also be allowed to fail occasionally, although not indefinitely or consistently. Evaluation of some applied research and most fundamental technology programs is more straightforward because the objectives are clearer and the causal chains more direct, although even here there are often surprises. For both science and technology, it takes astute and expert observers, and not bean counters, to tell how reasonable the gambles have been and how great the rewards should be, over appropriate periods.

Successful programs should be rewarded for achieving or sustaining world-class leadership. Unsuccessful ones should be eliminated, cut back, or reorganized. All programs should present compelling reasons for continuation or expansion. Criteria for success should suit the particular area of science or technology. Science intended only to advance understanding (e.g., archaeology or cosmology) will have different measures than mission-oriented fields (e.g., pharmacology or materials science) or fundamental technology (e.g., instrumentation or engineering). Individuals working in the fields are best able to judge value and craft appropriate measures.

47. Peter F. Drucker, "Really Reinventing Government," *The Atlantic Monthly* 275(2): 49, 1995.

48. One persistent theme of most reports on federal laboratories (note 10) is a strong need to free laboratories from "micromanagement" by federal agencies in Washington, D.C., and by Congress. This was a major concern of the Packard Report of 1983 (note 11). The Foster Report (1995) documents the number of task orders and NASA employees that oversee the Jet Propulsion Laboratory contract and judges them to be excessive. The Galvin Report (1995) cites this bureaucratic layering as among its top concerns. In the university setting, concerns have centered on the interpretation of Office of Management and Budget circulars A-110 and A-21, which set rules and accounting practices and in the judgment of many universities impose rigidities and induce inefficiencies, a concern addressed in the Federal Demonstration Project (see note 48).

49. The Federal Demonstration Project is described in annual reports of the Government-University-Industry-Research Roundtable (Washington, D.C.: National Academy of Sciences): *1993 Annual Report* (published April 1994), pp. 12-14, and *1994 Annual Report* (published 1995), pp. 9-10; and in the brochure "What Is the Federal Demonstration Project?" (August 1991), available from the Roundtable offices.

50. If functions of programs are shifted from federal responsibility, for example through block grants to states, the necessary R&D capacity must still be sustained. In transportation, state funding is channeled through a private national organization, whereas in public health, drug abuse, and health services most research remains funded by the federal government, with outreach to the states.

Part II

Supplements:
Background and Rationale

Supplement 1
The Evolution and Impact of Federal Government Support for R&D in Broad Outline

Today, the United States has the strongest research and development system in the world. Measured by the total amount of spending for or the number of persons employed in R&D,[1] the U.S. science and technology enterprise is the largest in the world. It is also the most successful. The U.S. garners the lion's share of the Nobel Prizes in physics, chemistry, medicine or physiology, and economics. Our nation sets the world standard for advanced education in nearly every field of science and engineering, and our high-technology firms are responsible for making and commercializing a substantial proportion of the important new technologies of our time.

In contrast, before World War II the United States was not as strong as the advanced countries of Europe in R&D. Private R&D spending was quite limited, university research was supported largely by private foundations and the states, and the federal government financed only about one-fifth of the nation's R&D.[2] Annual federal R&D expenditures at the eve of war in 1940 totaled under $70 million,[3] or about 1 percent of present-day expenditures, when adjusted for inflation.

Although the remarkable half-century interval from World War II to the present has been discussed in some detail elsewhere,[4] it is outlined here to provide some perspective on the historical processes that have shaped the current system of support for U.S. R&D. Study of the record reinforces appreciation of the depth and range of discoveries that continue to touch all aspects of our lives (see Box I.5 in Part I for a brief indication). It demonstrates that the federal role is essential in stimulating necessary new ideas and shows additional influences of federal government policy on U.S. science and technology. Strengths of the system will continue to serve national purposes well in the future.

The Contemporary Federal R&D Portfolio Resulted from Five Decades of Response to National Crises and Opportunities

Prior to World War II, most of the federal funds for R&D supported mission-oriented research in agriculture, national defense, and natural resources carried out by government employees in small government laboratories and experimental stations. Such R&D as was supported by the Army and Navy was done in military arsenals. Universities rarely sought federal funds for R&D, and many leading U.S. scientists obtained their advanced training in European universities. Industry received little government R&D money and looked to universities for technically trained staff and faculty consultants.

The evolution of the current system of support for U.S. science and technology can be outlined in terms of the following stages and events, among others:

• **Federal support of R&D grew remarkably in size and complexity during World War II.** Federal expenditures for R&D increased by an order of magnitude during World War II, and two important institutional innovations were introduced. First, large numbers of academic researchers were mobilized to work in their own institutions' laboratories on wartime R&D projects, whereas during World War I, scientists working on military projects had been made members of the military. Second, the R&D contract was devised as a mechanism to pay for private performance of work whose approach and outcome—in this case, R&D results—could not be specified precisely in advance. Importantly, the federal government agreed to compensate university and industry performers for the indirect or overhead costs of R&D done under grants and contracts, in addition to paying for direct expenses.

To carry out the vastly increased scale of R&D during World War II, major investments were made in research laboratories. New government laboratories were created and new administrative mechanisms were devised to oversee their work in the face of a shortage of government employees experienced in managing major R&D programs. A sense of mutual obligation emerged in which the R&D institutions could reasonably expect continued funding in return for producing quality efforts and results from government-financed programs.

- **Federal R&D support was consolidated in the immediate postwar period.** In his July 1945 report, *Science—The Endless Frontier*,[5] Vannevar Bush, who headed the U.S. wartime R&D effort, provided the intellectual rationale for federal support of both basic research and research related to national security, industry, and human health and welfare. He sketched a plan for a national research foundation, to be funded by the federal government and led by scientists from the private sector, that would support basic scientific research and education in areas related to medicine, the natural sciences, and new weapons. His plan contributed to legislation adopted in 1950 that established the National Science Foundation (NSF). By that time, however, the National Institutes of Health (NIH) had established its control over most health-related research, including university-based biomedical research and training; the Office of Naval Research (ONR) had taken on a major role in supporting academic research in the physical sciences; and the new Atomic Energy Commission had been assigned control of R&D on nuclear weapons and nuclear power. NSF's mission thus focused on supporting fundamental research and related educational activities, and its annual budget was less than $10 million until the late 1950s. In contrast, the NIH's annual budget, which had been less than $3 million at the end of the war, grew to more than $50 million by 1950.

- **The scope of federal R&D support grew modestly in the decade after World War II.** Several additional federal R&D efforts were launched during the late 1940s and early 1950s. Anxiety over the Cold War, and the loss in 1949 of the U.S. monopoly in nuclear weapons, led to expanded R&D programs in the Army and in the newly established Air Force, and to a continuing buildup in support for nuclear weapons R&D in the Atomic Energy Commission. On the civilian side, R&D programs were established or expanded in fields with direct practical importance, such as aeronautics technology, water desalinization, and atmospheric disturbances and weather. However, appropriations for these new civilian R&D efforts remained relatively limited through the mid-1950s.

- **Sputnik provided the impetus for a major expansion of federal support for R&D.** The launch of Sputnik by the Soviet Union in 1957 provoked national anxiety about a loss of U.S. technical superiority and led to immediate efforts to expand U.S. R&D, science and engineering education, and technology deployment. Within months, both the National Aeronautics and Space Administration (NASA) and the Advanced Research Projects Agency (ARPA) were established. NASA's core included the aeronautics programs of the National Advisory Committee on Aeronautics and some of the space activities of the Department of Defense (DOD); ARPA's purpose was to enable DOD to conduct advanced R&D to meet military needs and to ensure against future "technological surprise." Federal appropriations for R&D and for mathematics and science education in the NSF and other government agencies rose rapidly over the next decade, often at double-digit rates in real terms.

- **Growth of federal support for health research accelerated rapidly in the late 1950s.** During the early 1950s, growth in federal funding for health research slowed considerably from its torrid pace in the immediate postwar years. In the late 1950s, however, several factors converged to give renewed impetus to federal support for biomedical research: key congressional committees with responsibility for health-related research were chaired by powerful advocates of increased federal funding. Congress was appealed to by influential citizen advocates of increased funding for research to combat specific diseases. The calls for increased funding were supported by a strong NIH director, who could point to new scientific understanding of disease processes as the basis for anticipating medical breakthroughs. The result was the rapid growth of federal funding for health-related research that has continued nearly unabated to the present as new discoveries, and the rise of new diseases such as AIDS, have led to ever-greater commitments to biomedical research.

- **In the 1970s, new R&D-intensive agencies addressed environmental and energy issues.** Both the environmental movement and the energy crisis of the 1970s raised some doubts in American society about the wisdom of a national culture committed to consumption and economic growth, and led also to increased public and private spending on environmental and energy R&D. The energy agencies of the federal government were reorganized twice during the decade. In 1975, the Atomic Energy Commission was divided into the Energy Research and Development Administration and a new regulatory agency, the U.S. Nuclear Regulatory Commission. In 1977, the Energy Research and Development Administration and other federal energy-related activities were combined to form the Department of Energy (DOE), which was given major new responsibilities to fund energy-related R&D.

- **In the 1980s, the competitiveness challenge expanded the federal role in R&D and stimulated a new commitment to cooperation among industry, government, and universities in the conduct of R&D.** By the early 1980s, the industrialized world had largely recovered from the effects of World War II, and key Asian nations were devising new approaches to industrial production. The increasing challenges from competition abroad—in markets for traditional goods as well as a growing list of goods based on advanced technological capabilities—raised new questions regarding the role the federal government should play in assisting U.S. industry to develop and use new technology for competitive purposes. This topic remains under active debate today.

> **BOX II.1**
> **GOVERNMENT-UNIVERSITY-INDUSTRY COOPERATIVE R&D POLICIES**
>
> Government support of cooperative R&D involving firms, universities, and federal laboratories has roots in programs begun in the early 1960s—such as the Advanced Research Projects Agency's Materials Research Laboratories and the State Technical Services program in the Department of Commerce—and in the National Science Foundation's Industry-University Cooperative Research Centers program begun in the late 1970s. Such efforts expanded substantially in size and visibility with passage of the Stevenson-Wydler Technology Innovation Act in 1980. The act also made technology transfer to industry and states a mission of all federal laboratories. The Federal Technology Transfer Act of 1986 later authorized government-operated federal laboratories to enter into cooperative research and development agreements (CRADAs) with companies and consortia of companies to pursue projects of mutual interest. In the early days of CRADAs, no money was exchanged between the laboratory and the participating firms, and the agencies and their laboratories did not have specific budgets to support their work with firms. More recently, as the contractor-operated federal laboratories were authorized by the National Competitiveness Technology Transfer Act of 1989 to enter into CRADAs, the Department of Energy, which owns most of these laboratories, has set aside funds in its defense programs and energy research budgets to fund, on a competitive basis, laboratory R&D that contributes to specific CRADAs.
>
> The Small Business Innovation Development Act of 1982 required all federal agencies that spend a significant amount on R&D to set aside a small proportion of those funds to support R&D projects of interest to them at small businesses on a competitive basis. These Small Business Innovation Research grants are intended to assist small firms in developing new products to serve a federal requirement and/or a commercial market. In 1985, NSF was given a budget to fund engineering research centers at universities, with the proviso that the award of government funds was contingent on industrial support for those centers. This program was later expanded to support science and technology centers as well on a similar basis.
>
> The Omnibus Trade and Competitiveness Act of 1988 authorized the National Institute of Standards and Technology (NIST) to establish an Advanced Technology Program of competitive awards to firms and consortia of firms on a matching basis to support early-stage, generic technology development projects. The same act authorized what has become the Manufacturing Extension Partnerships program in NIST, which provides grants to nonprofit consortia and state and local governments for transfer of technology and technical assistance to manufacturing firms, with an emphasis on small- and medium-sized firms.
>
> An amendment to the Defense Authorization Act for Fiscal Year 1993 established the authority for the Department of Defense, in cooperation with other federal agencies, to fund a variety of technology development, technology deployment, and technical education and training activities at firms, consortia of firms, and nonprofit organizations. This authority was used to create the Technology Reinvestment Program in 1993. Led by the Advanced Research Projects Agency, the Technology Reinvestment Program involves the Departments of Commerce, Defense, Energy, and Transportation, as well as the National Science Foundation and the National Aeronautics and Space Administration.
>
> A number of these programs are under considerable scrutiny by the 104th Congress, and some of them face elimination or sharp budget reductions.

During the 1980s and early 1990s, several programs were initiated to provide financial and other incentives for industrial R&D and for industrially related R&D conducted at universities or federal laboratories (see Box II.1). These included the Small Business Innovation Research program, the NSF Engineering Research Centers, and the Advanced Technology Program and Manufacturing Extension Partnerships at the Department of Commerce. In addition, federal policy changes enabled the creation of the cooperative research and development agreement, or CRADA, a mechanism for joint R&D involving companies and federal laboratories.

- **Throughout the five decades following World War II, federal funds for R&D were reduced substantially in only one period.** The costs of the Vietnam War squeezed nondefense R&D along with other nondefense discretionary spending. From 1966 to 1975, federal support for nondefense R&D dropped nearly 22 percent in real terms. The successful conclusion of NASA's Apollo program contributed to the decline in federal R&D funding during that period, as did skepticism about the value of advanced technology that was engendered by the Vietnam War and the contemporaneous environmental movement.

Since the mid-1980s, the continuing struggle to control federal budget deficits has put increasing pressure on federal R&D funding. R&D programs have had to compete for money more directly with other federal activities and have also been affected by the various mechanisms adopted to enforce budget deficit reduction, including the Balanced Budget and Emergency Deficit Control Act of 1985 (commonly known as the Gramm-Rudman-Hollings Act) and its amendments as well as the Budget Enforcement Act of 1990.

Budgetary pressure on federal R&D spending is intense today. Federal funds previously appropriated to support R&D during Fiscal Year 1995 have been cut (rescinded) by nearly $2 billion. Furthermore, much larger cuts in federal R&D funding are slated for Fiscal Year 1996, and pressures on federal discretionary spending make further cuts in future years likely.

Key Roles of the Federal Government in U.S. Research and Development

In keeping with national aspirations and the practice of governments of all advanced nations, the federal government provides a substantial proportion of the direct financing for R&D done in this country, and it also offers incentives to private interests to support R&D. Many other federal policies affect the performance of R&D and the use of its results—some policies stimulate such activity, while others create barriers to it.

The federal government invests in building and strengthening the research and development essential to pursuing a variety of national goals.

Much of the federal science and technology investment is intended to help build the base of scientific and technical knowledge and expertise used by government and industry to address important national goals, such as national defense, space exploration, economic growth, and protection of public health and the environment. The federal government has assumed a central responsibility for supporting graduate education in science and engineering because of its critical importance to the continuing vitality of the nation's innovation system. Most of this support is provided by the funding of R&D at universities, which offers students the opportunity to carry out cutting-edge research as an integral part of their education.

Indirect federal financial support encourages a climate of opportunity for R&D in the United States.

In addition to granting funds directly to performers of R&D, the federal government creates incentives for private spending on R&D in industry and academic institutions:

- Since its inception in 1790, the U.S. patent system, for example, has provided an incentive to inventors to develop and to disclose, use, and profit from their inventions.

- Since 1954, industry has been able to deduct the full costs of R&D from income before taxes in the year in which they were incurred, while depreciating the costs of facilities and major equipment. Since passage of the Economic Recovery Tax Act of 1981, a series of special tax credits have been offered to firms that increase their R&D spending above previous levels. Individuals and corporations that make charitable contributions in support of research in educational institutions also are eligible for tax savings.

- The Stevenson-Wydler Technology Innovation Act of 1980 opened the federal laboratories to industry, making available not only specialized and unique facilities, but also opportunities for R&D partnerships with joint funding and the use of federally developed technology for profit-making ventures. That same year, Congress passed the Bayh-Dole Act, which conferred ownership of patent rights to universities, small businesses, and nonprofit organizations, thus providing a strong incentive for commercial development. In 1984, the National Cooperative Research Act amended the antitrust statutes to facilitate cooperative R&D among competing firms.

- With increasing frequency, the federal government has cost-shared with firms and consortia to underwrite precompetitive technology development projects in such areas as manufacturing technology or technology with a strong potential for application in both defense and commercial arenas (so-called dual-use technology).

- By formally and informally identifying areas of technological opportunity and by convening experts from a variety of organizations to address technical topics, government leadership helps initiate cooperative R&D ventures that otherwise might not be arranged by competing firms.

Many other federal policies and programs have indirect effects that can foster or impede innovation and affect the environment for R&D.

Policies in many areas can have dramatic, if indirect, effects on private spending on research and development and, hence, innovation. For example, tax code provisions of the kind mentioned above, such as accelerated depreciation, investment tax credits, and capital gains preferences, can reduce the corporate cost of capital for R&D investments and increase the supply of risk capital to commercialize new technologies. Trade policy can open new markets for high-technology goods. Regulation is centrally important for new drugs and agricultural products.

Some public policies, however, can hinder the conduct of R&D in universities, industry, and other private institutions, even though that is not their aim. Adopted in pursuit of important societal purposes, some, for example, raise the direct and indirect costs of conducting R&D. Private performers of R&D must comply with a host of laws and regulations intended to affect conduct generally, in such areas as antitrust, labor relations, equal opportunity, consumer safety, and environmental protection. Nongovernmental recipients of public R&D funds must comply with additional rules and regulations regarding the procurement process, financial accountability, nondiscrimination and affirmative action, preferences for small and minority-owned businesses, "Buy American" requirements, maintaining a drug-free workplace, and so on.

Results of 50 Years of Federal R&D Support

Investment in R&D has become an essential element of contemporary governance.

A history of successful experiences in mobilizing scientific and technical resources to meet important national needs has contributed to a sense of confidence that U.S. scientific and technical institutions can rise to nearly any occasion and help address important national problems with dispatch. Congress, the Executive Branch, and the American people have come to believe that investment in R&D is a cost-effective mechanism for responding to important national needs. R&D helps ensure our national security, strengthens the performance of our economy, and enhances our quality of life.

The United States is not alone in this belief—during the twentieth century every industrialized country has made major investments in the foundations of its scientific and technological capabilities through support for R&D and related activities. In fact, support for R&D is now one of the primary tools used by modern governments everywhere to achieve public purposes.

The breadth of the federal investments in R&D provides the scientific and technical capital to respond to new opportunities and crises, which often are unexpected and sometimes are urgent.

U.S. strength in a wide range of fields has enabled both creative and pragmatic problem solving on diverse fronts: rapid understanding of the factors related to the onset of AIDS, responses to new forms of warfare, and identification of major environmental problems such as losses in stratospheric ozone.

Diversity, both in funding sources and in the institutions that do the work, is a great strength of our national science and technology enterprise.

Research and development supported by ONR, NSF, NASA, and the U.S. Geological Survey has led to a revolution in our understanding of Earth's structure, its resources, and the impact of geological forces. Similarly, U.S. strength in information technology has been fostered through the work of DOD, NSF, DOE, and other agencies. Often several agencies have collaborated to create a successful program.

The support and policies of DOD and NSF, for example, led to the creation of the Internet; several agencies have contributed to the U.S. strength in the optical sciences.

At the same time, one agency may be the primary, if not sole, patron of a field of national importance; for example, DOE is the largest supporter of academic research in nuclear physics. DOD's support of computer science and engineering and materials science and engineering enabled the creation of Silicon Valley, and support by NIH facilitated the emergence of modern biotechnology.

The federal budget allocation process allows for this diversity of approach in which budgeting is handled mainly by agencies who know well the purpose and content of R&D projects and need their results. Budget decisions are thus specific to programs rather than generalized and across the board, and good science can find sustenance wherever it first arises.

Stable and thoughtful research investments can contribute to controlling federal costs.

Continuing technological superiority enables the United States to maintain a reduced but highly effective military force without compromising national security; new nondestructive testing techniques reduce the costs of maintaining highways; and information technologies help federal agencies, such as the Social Security Administration and the Internal Revenue Service, control the costs of serving very large populations. Through prevention of disease and development of new therapies, biomedical research has the potential to reduce significantly the costs of disease, injury, and health care.

Major advances in technology often are based on research whose eventual outcomes and applications could not have been predicted.

The de facto postwar policy of "poised to pounce"—that is, the readiness to respond made possible with support across a wide spectrum of the sciences, complemented by funding targeted to particular opportunities and priorities as they become apparent—has worked. Major advances have come from unexpected sources. For example, fundamental work on atomic clocks led to the concept and development of the global positioning system (Box II.2); work on the microwave spectrum of ammonia enabled the development of lasers; and studies of magnetic moments and nuclear spin were the basis for the development of magnetic resonance imaging and dramatic new forms of medical diagnosis. Research on the genetics of bacterial viruses and harmless bacteria that live in the human gut contributed to advances in biotechnology, and the study of large biological molecules by x-ray diffraction has greatly aided the effort to design new drugs.

Decades of separate lines of work in biology, psychology, linguistics, and anatomy have converged to create neuroscience, in which fundamental work holds the potential for enormous rewards—from better treatments for mental illnesses to improved ways of teaching and learning to the design of radical new computer

> ## Box II.2
> ### Origins of the Global Positioning System
>
> The global positioning system (GPS), a satellite-based system enabling remarkably precise pinpointing of one's location on Earth, is a contemporary product of a diverse R&D system. GPS evolved from postwar work on atomic clocks to test aspects of general relativity theory. Their possible value for navigation was recognized by the military, which provided years of "patient federal capital" to mature the technology. While the military's primary interest in what was to become GPS was to improve the delivery of tactical weapons and to reverse the proliferation of costly new navigation systems, its civilian potential was seen at the outset; that is, early in its development GPS was recognized as a potential dual-use technology, and in fact the commercial GPS market now overshadows military demand.[1]
>
> Several military programs involved in what was to become GPS coalesced in 1972, when the Air Force was given responsibility for developing a navigation system for all military services as well as civilian users. Concurrently, technologies essential to GPS, including satellites and microelectronics, also were being developed. Experimental GPS satellites were launched in 1978, and proof that GPS could be used for locating one's place on Earth soon followed. Eighteen GPS satellites were launched by the United States by 1990. Today's system consists of 24 satellites, each carrying up to four atomic clocks that provide timing and ranging signals. A GPS receiver decodes the signals to determine and display their latitude, longitude, and altitude. Differential GPS is the most widely used method for augmenting basic GPS signals and now yields centimeter accuracies over distances of several kilometers. That translates into what is already an incredible array of applications, such as demonstrating new systems for landing aircraft in bad weather (i.e., a fully automatic CAT II aircraft landing); robotic plowing, planting, and fertilizing of fields; monitoring train locations; and tracking and cleaning up oil spills. The 1995 global GPS market is estimated at $2.3 billion today and is projected to reach $11.6 billion by 2000.[2] Civil production of GPS units is now more than 70,000 per month.
>
> Secretary of Defense William J. Perry recently commented that the "GPS system ... was the key to being able to find and rescue Capt. Scott O'Grady [the Air Force pilot shot down June 2 and rescued June 8, 1995] and pull him out of Bosnia....That whole operation would not have been possible except for the fact that Capt. O'Grady had a little GPS receiver on his wrist and the incoming helicopters had a receiver....The consequence—they landed essentially at his feet, and the total time on the ground was less than two minutes. If they had had to spend a half hour or so searching for him, the results could have been very different."[3]
>
> ---
>
> [1] National Academy of Public Administration, *The Global Positioning System: Charting the Future* (Washington, D.C.: National Academy of Public Administration, 1995), pp. 5, 14.
> [2] National Academy of Public Administration, *The Global Positioning System*, 1995, p. 15.
> [3] Prepared remarks of Secretary of Defense William J. Perry to the Economics Engineering Systems Department graduating class, Stanford University, Stanford, Calif., June 18, 1995.

architectures. The Decade of the Brain, a 10-year federal commitment to exploit the advances of many facets of brain research conducted through multiple departments and agencies, is inherently interdisciplinary. The program has several specific goals that encompass diverse areas of science, and it incorporates a wide range of technologies used in brain imaging, molecular genetics, and computer analysis of complex biological structures.[6]

Scientists and engineers whose education and training have included opportunities to conduct research in universities have served the nation well.

Linking federally funded research and development to the education of scientists and engineers has powerfully enhanced both. Universities are the core strength of the U.S. R&D system. They are by far the most important source of men and women educated and trained in advanced science and engineering. Such people, as they establish their own university careers, join industry, or start their own companies, are the most effective and efficient agents of technology transfer. Experience demonstrates that the excellence of the next generation of researchers and leaders depends directly on the excellence of graduate education that includes first-hand participation in innovative research and development. Over the last several decades, federal support for academic research has been crucial to maintaining that linkage.

The existing U.S. research and development system works well in periods of continued expansion in missions and funding but is not as appropriate in periods of static or declining budgets.

The U.S. R&D system is largely the creation of a period of unprecedented growth in private economic activity and government programs in the United States. The current federal R&D budgeting process evolved to accommodate new missions, and the performing institutions grew to meet the challenge of growing federal expectations and increased appropriations. Flexibility was achieved mainly by building new structures, not by devising means to change old ones. The research and development system is conditioned on growth and is now challenged by the new environment that requires downsizing of both missions and budgets.

Scientists and engineers can respond fairly quickly to new research opportunities and changes in funding emphases. Similar flexibility is more difficult for large research institutions to manage.

The U.S. research and development system is changing in response to changing national circumstances. DOD has combined a number of its R&D facilities and has closed others. Many major firms have refocused their corporate long-range R&D laboratories on more immediate business needs and opportunities. Such changes reflect shifts in the federal research portfolio, which has changed dramatically over the decades since the onset of World War II, both in launching new programs, such as planetary exploration, and in reducing others, such as the breeder reactor program. But flexibility of project funding in some areas has not been matched by flexibility in large R&D institutions and facilities. The nation now carries an excess of facilities, many established during World War II and the Cold War, whose missions may no longer be appropriate or whose programs may not be as competitive as others. Their continued support will detract from more effective or more important programs, inhibiting a vigorous research enterprise in an era of limited resources.

Supplement 2
Federal Funds for R&D and FS&T

Distribution of Federal Funds for R&D as Currently Reported

At present, the federal government invests about $70 billion annually to finance the conduct of R&D in industry, federal laboratories, academia, and independent research organizations. Of the nearly $70 billion spent on R&D in Fiscal Year 1994, federal science and technology (FS&T), as defined by the committee (Box II.3), received between $35 billion and $40 billion, while the remaining portion was devoted to demonstration, testing, and evaluation of major systems.

In Fiscal Year 1994, about 45 percent of the federal R&D funds went to industry, 25 percent to the federal government's own laboratories (not including FFRDCs), 17 percent to institutions of higher education, 8 percent to FFRDCs, and about 5 percent to other nonprofit or nonfederal research institutions.[1]

Based on standard current definitions, the federal government funds about 36 percent of all R&D in the United States.[2] In recent years, the federal government has supplied about 60 percent of the funds that support R&D in educational institutions, almost 20 percent of the funds for R&D in industry, and essentially all of the support for R&D in federal laboratories.[3] Thus, it is apparent that federal funding has been essential to R&D performance in all three sectors.

The Usefulness of Thinking About a Federal R&D "Portfolio"

The federal government invests in a highly diversified portfolio of R&D in many disciplines and for many purposes. This portfolio includes programs and projects with widely different expected risks and pay-off horizons, is the responsibility of many federal departments and agencies, and is pursued in a variety of institutions. No single decision-making model is appropriate to investments in all elements of the portfolio; in fact, the different elements in the portfolio are established in quite different ways and at different levels. The federal government has not worked with a federal "budget" as such; instead, total annual spending on R&D by the federal government has resulted from the aggregation of the results of decisions made by separately compiling the budgets of the diverse departments and agencies.

The Nature of the Contemporary Federal R&D and FS&T Portfolios

In this section, the committee summarizes its understanding of the salient features of the contemporary federal R&D and FS&T portfolios. The R&D data are taken largely from standard statistical sources, and, unless otherwise noted, are presented using the categories and definitions employed by the Division of Science Resources Studies of the National Science Foundation. The FS&T data were developed by the committee, and their derivation is discussed in Box II.3. Several questions about the FS&T budget concept are addressed in Box II.4.

Box II.3
The Federal Science and Technology (FS&T) Budget Concept

For policymaking purposes, the key feature of research and development activities is their investment nature. Increasing the stock of knowledge and devising new ways to apply that knowledge are major sources of future growth and security. Research and development in the federal budget are not current-consumption items; decisions on federal support for research and development should take into account their future contributions to better health, greater military and economic security, quality of life, and human knowledge. It is especially important to factor in the future investment nature of research and development when budgets are being determined. Federal policymakers will want to sustain future economic growth, in part because it is an important way to address budget deficits in the long term.

The committee understands fully that there is great uncertainty in research and development investments. The processes leading to commercially viable and socially useful technologies are complex and involve substantial non-R&D factors. That makes investments in research and development necessary but not sufficient for technological progress. The uncertainty of where discoveries will be made and which of them will have practical uses underlies the committee's recommendation that the United States perform at the world-class level, if not lead the world outright, in all areas of science and technology (see Recommendation 4 in Part I of this report).[1]

As currently reported, federal spending for research and development totals approximately $70 billion a year. However, nearly half of traditional federal research and development spending involves initial production, maintenance, and upgrading of large-scale weapons and space systems at the Department of Defense, Department of Energy, and National Aeronautics and Space Administration. Those activities are neither long-term investments in new knowledge nor investments in creating substantially new applications. If they were excluded, the research and development investment budget—called the *federal science and technology* (FS&T) budget in this report—would be between $35 billion and $40 billion annually.

The Department of Defense, which has by far the largest budget for research and development (nearly half of the $69.6 billion obligated by all federal agencies for research and development in Fiscal Year 1994), has already begun to distinguish between "science and technology" and "systems development" in its research and development budget (see Table II.1). The Department of Defense's definition of science and technology, which is essentially the same as that used for FS&T in this report, includes the first three of the seven research and development categories that the Department of Defense uses; systems development corresponds to the other four Department of Defense categories for research and development (see Table II.1). In Fiscal Year 1994, approximately $24.6 billion in research and development activities supported by the Department of Defense fell outside what this report identifies as federal science and technology (FS&T).[2]

Unlike the Department of Defense, the Department of Energy and the National Aeronautics and Space Administration do not break out the development portions of their research and development budgets by subcategories, and it is more difficult to determine how much of the research and development at those agencies should be classified as FS&T and how much excluded.[3] The Office of the Director of Defense Research and Engineering at the Department of Defense estimated that in Fiscal Year 1993, about $5.1 billion of NASA's research and development budget of $8.0 billion—and about $5.0 billion of DOE's research and development budget of $6.3 billion—was equivalent to DOD R&D categories 6.1 through 6.3A and thus should be included in FS&T.[4]

If the FS&T estimates for Fiscal Year 1994 for DOD ($8 billion), DOE ($5 billion), and NASA ($6 billion) are added to the research and development totals for the other agencies ($19 billion), the approximate total for FS&T is $37.6 billion. Because that number incorporates some rough estimates, especially for DOE and NASA, the text of this report uses the range estimate of $35 billion to $40 billion for FS&T.

---

[1] See also COSEPUP (National Academy of Sciences, National Academy of Engineering, Institute of Medicine), *Science, Technology, and the Federal Government: National Goals for a New Era* (Washington, D.C.: National Academy Press, 1993); Ralph E. Gomory, "The Known, the Unknown, and the Unknowable," *Scientific American* 272 (June 1995): 120.

[2] DOD is currently working with the Division of Science Resources Studies at the National Science Foundation to report its R&D spending in two categories, "science and technology" and "systems development." That exercise should result in more precise estimates of FS&T spending by DOD and how it is distributed among performing institutions.

[3] It is possible that some R&D activities in agencies other than DOD, DOE, and NASA would not qualify to be FS&T, but the amount is probably negligible. In any case, the other agencies account for a very small portion of federal expenditures on development—less than 6 percent ($2.4 billion) in Fiscal Year 1994.

[4] The figures were presented by Dr. Anita K. Jones, director of Defense Research and Engineering, at the January 1995 meeting of the committee. They were rough "guesstimates" made on the basis of telephone calls from Office of the Director of Defense Research and Engineering staff to contacts at DOE and NASA.

TABLE II.1 Department of Defense R&D Budget (dollars in thousands)

		FY 1994	FY 1995	FY 1996
Science and Technology				
R&D Category				
6.1	Basic Research	$1,167,211	$1,227,021	$1,213,918
6.2	Exploratory Development	2,760,676	3,069,940	2,816,061
6.3A	Advanced Development	3,898,100	4,339,424	3,796,157
	TOTAL S&T	$7,825,987	$8,636,385	$7,826,136
Systems Development				
R&D Category				
6.3B	Demonstration/Validation	$ 2,696,592	$ 4,324,990	$ 4,229,027
6.4	Engineering & Manufacturing Development	7,334,269	8,930,372	8,759,104
6.5	Management Support	3,367,685	3,435,590	3,305,088
6.6	Operational System Development	11,241,890	10,187,818	10,212,598
	TOTAL Systems Development	$24,640,436	$26,878,770	$26,505,817
	TOTAL DOD R&D	$32,466,423	$35,515,155	$34,331,953

NOTE: Adapted from Department of Defense data provided by R. Tuohy through private correspondence.

Box II.4
Using the FS&T Budget Concept: Questions and Answers

Why adopt a new budgeting concept for R&D policymaking?

In a period of severe constraints on the federal budget and reduced allocations for R&D, it is especially important to focus on the investment aspects of federal science and technology. The part of the R&D budget that supports science and key enabling technologies must be identified and considered in overall terms by Congress and Executive Branch policymakers. It is FS&T that expands the stock of knowledge about the physical, biological, and social world and finds new ways to use that knowledge productively. Decision making on the rest of the R&D budget concerns testing and evaluation of large technical systems prior to production, and their subsequent modernization, and thus rests on different and shorter-term considerations than do the criteria for allocating funds for FS&T as defined in this report.

Will the new FS&T concept confuse people familiar with the conventional R&D numbers that have been used since the early 1960s?

Although continuity and comparability in data series are useful for policy analysts, it is more important for those making allocation decisions to have data that measure the right things. The usefulness of FS&T data and the increased effectiveness of budgeting based on them will more than outweigh the costs of implementing and learning how to interpret the new data series. In any case, OMB and NSF can continue to collect and report the traditional R&D totals, of which FS&T data are a subset. That approach is similar to the one now being taken by NSF and DOD in collecting data on the science and technology and the systems development parts of R&D at DOD (see Table II.1).

Do available data allow for departments, agencies, OSTP, OMB, and Congress to use the FS&T budgeting concept practically and unambiguously?

To implement the FS&T budget concept fully, some new data will have to be collected and some new interpretations of existing data must be made by some agencies. However, the agency most affected by the new approach—DOD—already tracks its R&D activities in a way that feeds directly into FS&T estimates. Making such determinations in DOE, NASA, and perhaps other agencies should be relatively straightforward after experimentation with one or two years' budgets. Some funding in higher categories may support the science and technology base. Independent R&D funds in federal procurement contracts (which are no longer reported fully) and some facilities and infrastructure elements may contain items that intuitively belong in FS&T. The Internet grew out of one such account, for example. Over time, the FS&T concept and the data it generates will become a normal part of the budget process, and the current imprecision signified by the committee's range estimate of $35 billion to $40 billion annually will narrow.

Why not just use trends in the basic research or total research (basic and applied) subcategories as a budget indicator for the science and engineering enterprise rather than invent a new category?

The strength of the FS&T budget concept is that it corresponds to the set of research and technology development activities typically conducted in the science and engineering departments of U.S. research universities, many of the federal laboratories and FFRDCs, and some private firms. Those institutions conduct a rich, interactive mix of investigations aimed at discovering new knowledge of fundamental phenomena and their applications. Just looking at basic research or even basic and applied research is too narrow for federal policymaking.

If development is a continuum, isn't excluding a part of it as being too tied to the acquisition or upgrading of specific systems merely arbitrary?

Applying any definitional categories, whether the familiar R&D or new FS&T ones, to complex reality involves some arbitrariness. The only advantage of retaining the old definition of development for DOD, DOE, and NASA is that long usage has probably made categorization decisions more consistent (but not necessarily more valid). The committee believes that FS&T corresponds more closely to the common-sense definition of R&D that most people hold, and its adoption will not lead to serious or long-term inconsistencies or confusion. R&D activities beyond FS&T typically spend most of their financial and human resources on systems-operation-type activities rather than the pursuit of new knowledge and novel applications.

Does using the smaller base give those who want to protect the funding of fundamental science and technology less to trade off in a period of serious budget cutting?

The report points out that such trade-offs are not—cannot be—made under the current budget structure, because the current R&D budget is not actually used for budgeting purposes. It is totaled after the fact and is based on a series of trade-offs made at the agency level or lower. Specifically, the $25 billion in DOD R&D that is separate from FS&T cannot be reallocated to other areas even within DOD, let alone to other parts of the federal budget. After lengthy debate on this issue, the committee concluded that supporters of a strong science and technology enterprise in the United States are better off defending the smaller FS&T budget than retaining the larger traditional R&D number in hopes of capturing some of the funding for such systems engineering and operational support as upgrading the Navy's F-14s. The greater problem may be protecting the FS&T base from the major cutbacks in systems approaching the full procurement stage.

Will use of the FS&T budget concept throw off international comparisons?

The committee did not study the issue in any depth but has the impression that only a few other countries' budgets for science and technology include systems development for national defense of the kind that DOD does, and so the FS&T number is a more accurate basis for international comparisons than is the currently reported number for federal R&D. The important thing is to use the right number, one that truly measures R&D and is consistent with the numbers reported by other nations. More work will be needed to clarify the meaning of international science and technology budget comparisons.

What fields of science and technology are included in the FS&T base?

The FS&T base is defined as work intended mainly to produce new knowledge or new technology, and so it includes the full range of fields in science and engineering: the life sciences, physical sciences, environmental or geosciences, mathematical and computer sciences, psychology, social sciences, and engineering. These are the same fields included by NSF and OMB in calculating federal R&D. The FS&T base also contributes to a broad range of national programs beyond the well-known ones of health, defense, agriculture, energy, space, and fundamental disciplinary research. Work in the FS&T base is also conducted to improve transportation systems and other types of public works infrastructure, environmental remediation, work education programs, criminal justice, standards and measures, research background for regulatory actions, and many other areas of public concern.

Box II.4 continues on next page.

> **BOX II.4 CONTINUED**
>
> *What method was used to estimate the levels of funding for the FS&T base shown in the figures in this report?*
>
> A number of assumptions and sources of data were used to approximate the levels of funding for the FS&T base (they are detailed in the caption for each figure). The general approach was to subtract the advanced systems development funding of DOD, NASA, and DOE from total federal R&D spending as currently reported:
>
> - Funding of research by all federal agencies was included;
> - Funding of development by all federal agencies *except the Department of Defense, National Aeronautics and Space Administration, and Department of Energy* was included;
> - Funding of what DOD calls Research Category 6.3A was included, as reported by the Office of the Director of Defense Research and Engineering (ODDR&E). Thus, funding of categories 6.3B through 6.6 was not included; and
> - Finally, and most roughly, funding of the equivalent of 6.3A-type activities by NASA and DOE was included (based on estimates for FY 1993 made by ODDR&E).
>
> The procedure outlined above yields an estimate of $37.6 billion for the FS&T base in Fiscal Year 1994. Because that number is based on a series of approximations and extrapolations, the range of $35 billion to $40 billion is used in this report. The point estimate of $37.6 billion is used for illustration in the accompanying figures, with similar estimates for other years (see Box II.3).
>
> ---
>
> [1]These fields are listed and defined in National Science Foundation, *Federal Funds for Research and Development: FY 1992, 1993, and 1994*, NSF 94-328 (Arlington, Va.: National Science Foundation, 1995), pp. 6-9.

Federal R&D supports both a core of FS&T and a set of activities closer to production or application.

Most federal departments and agencies report their total investments in R&D within three categories: basic research, applied research, and development. However, for some agencies—in particular DOD, DOE, and NASA—R&D expenditures include the costs of activities that in other agencies or in the private sector might be considered as outside the scope of R&D, including engineering development, upgrades and modernization, testing and evaluation, and the like. As discussed in Part I of this report, the committee focuses on the FS&T investments of the federal departments and agencies. For most of them, FS&T is identical to R&D. For DOD, DOE, and NASA, however, the committee excludes demonstration, testing, and evaluation of existing technologies from FS&T. For Fiscal Year 1994, the committee estimates that total federal R&D funding was approximately $70 billion, while FS&T funding was between $35 billion and $40 billion.

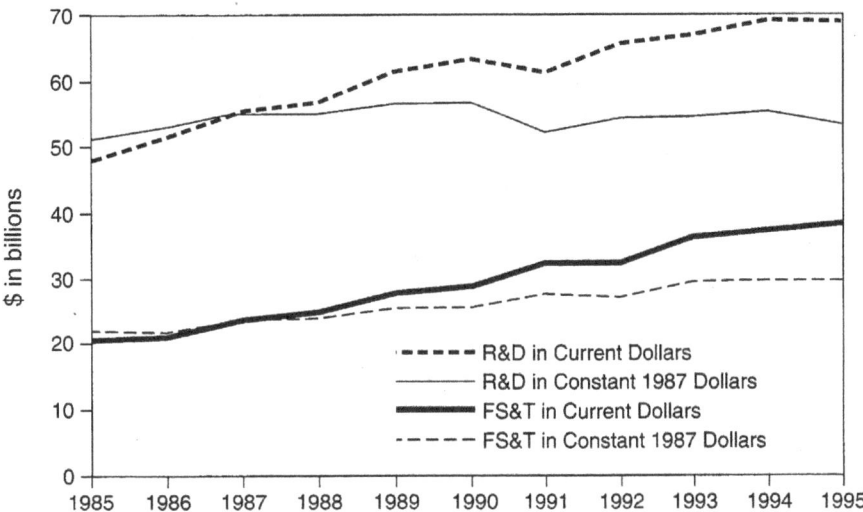

FIGURE II.1 Trends in federal support of R&D and FS&T, Fiscal Year 1994.

SOURCE: Data on federal R&D from Table C-93a, NSF, *Federal Funds for Research and Development: Fiscal Years 1993, 1994, and 1995,* NSF 95-334 (Arlington, Va.: NSF/Division of Science Resources Studies, forthcoming). The data for FY 1985 through FY 1993 are actual obligations; those for FY 1994 and FY 1995 were estimated by the R&D agencies. The GDP implicit price deflators (1987 = 100) were taken from Table B-1, NSF, *National Patterns of R&D Resources: 1994* (NSF/Division of Science Resources Studies, 1995), p. 8. FS&T numbers were derived from agency R&D budgets by subtracting spending for DOD research categories 6.3b through 6.6 and spending for equivalent activities at NASA and DOE in 1993, as estimated by the Office of the Director of Defense Research and Engineering, and extrapolated to 1994.

The federal R&D (and FS&T) portfolio is complex and diverse.

Figure II.1 shows the trends over the last decade in federal R&D and FS&T funding in both current- and constant-dollar terms.[4] While the current-dollar curve suggests a slow, steady rise in federal R&D spending up until Fiscal Year 1994, the constant-dollar curve shows that total federal R&D spending peaked in Fiscal Year 1990. The downturn from 1994 to 1995 is actually larger than indicated in the figure, because nearly $2 billion in Fiscal Year 1995 budget authority has been subsequently rescinded. The President's budget for Fiscal Year 1996 calls for cuts of about 20 percent in real terms over the period from 1996 to 2000, and congressional spending plans call for even larger reductions in R&D—33 percent in real terms by Fiscal Year 2002, according to the budget resolution of June 1995.[5]

FS&T has shown a somewhat different pattern, owing to the subtraction from R&D of the very rapidly changing and large amounts of spending in non-FS&T programs. In Fiscal Year 1987 dollars, FS&T funding grew steadily from 1985 through 1993 and has been essentially constant in 1994 and 1995.

Figure II.2 shows trends in the ratios of federal support for R&D and FS&T to gross domestic product (GDP). The federal government has recently invested the equivalent of about 1 percent of the GDP in R&D, although the ratio has been slowly declining for some time, from nearly 1.5 percent 25 years ago. The proportion of GDP corresponding to FS&T has been growing slowly and is now in the neighborhood of 0.5 to 0.6 percent.

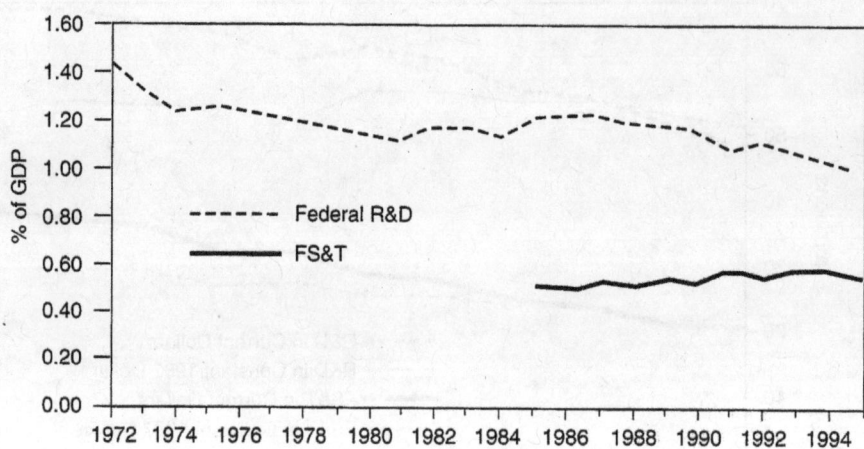

FIGURE II.2 Trends in federal R&D and FS&T spending as a percentage of GDP, Fiscal Year 1994.

SOURCE: Federal R&D and FS&T figures are from the sources cited for Figure II.1; GDP data are from Table B-1, National Science Foundation, *National Patterns of R&D Resources: 1994* (Arlington, Va.: NSF/Division of Science Resources Studies, 1995), p. 8.

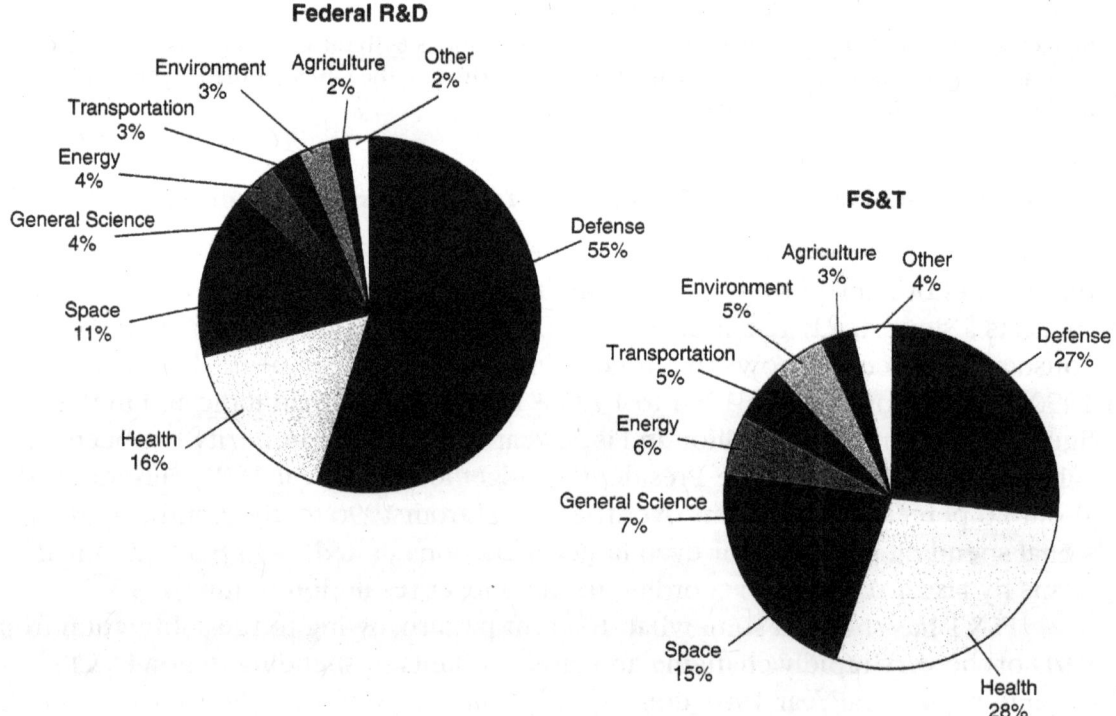

FIGURE II.3 Federal R&D and FS&T funding by national goal, Fiscal Year 1994.

SOURCE: Data on federal R&D are from Table 4, National Science Foundation, *Federal Funding by Budget Function: Fiscal Years 1993-95* (Arlington, Va.: NSF/Division of Science Resources Studies, forthcoming). FS&T data were derived by substituting FS&T funding by DOD, DOE, and NASA from Figure II.1 for National Defense, Energy, and Space Research and Technology R&D totals (this exercise involves making a somewhat arbitrary division of DOE FS&T between national defense activities (atomic energy) and energy activities).

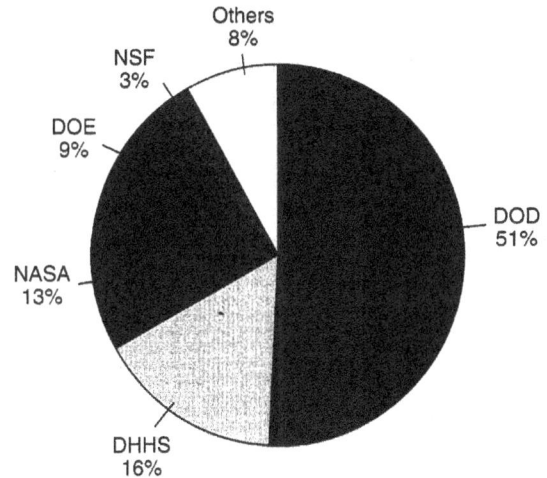

FIGURE II.4 Distribution of R&D funds among the agencies.

SOURCE: Data are from the American Association for the Advancement of Science, unpublished tables of federal R&D funding by budget function and agency, Fiscal Years 1994 through 1996, provided by Kei Koizumi, Directorate for Science and Policy Programs, AAAS, September 26, 1995.

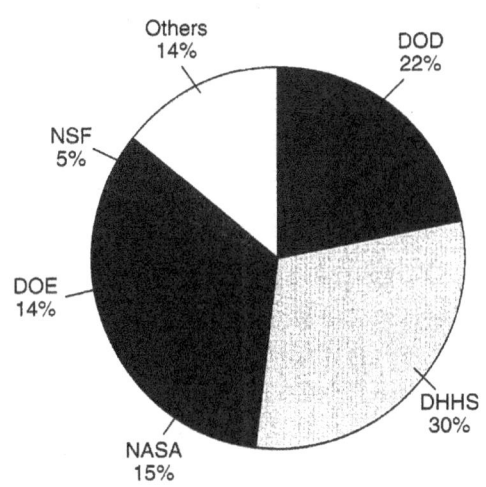

FIGURE II.5 Distribution of FS&T funds among the agencies.

SOURCE: Data as for Figure II.4, modified as noted in Figure II.1.

The federal government supports R&D in the pursuit of diverse national goals and objectives. Federal spending for R&D is heavily focused on defense, health, space, and energy, as indicated in Figure II.3. FS&T funding is less heavily focused on defense, and a greater portion is devoted to health and other topics. Reflecting the diverse goals of federal R&D spending, most federal departments and agencies support at least some R&D, as illustrated in Figure II.4. Figure II.5 shows FS&T funding as allocated among the agencies.

Federally supported R&D is performed in diverse institutions, including government laboratories, industry, academic institutions, and independent R&D organizations (see Boxes II.5 and II.6). Figure II.6 shows the breakdown of federal R&D spending among the different categories of performing institutions for Fiscal Year 1994. Note that industry is by far the largest performer of federally funded R&D, followed by government laboratories and then academia, with other nonprofit institutions playing the smallest role. As Figure II.7 indicates, the largest proportion of FS&T is performed by government-owned, government-operated laboratories; academic institutions are the second largest performers; and industry is in third place.

Box II.5
Categories of R&D Performers

Thousands of institutions in the United States conduct R&D, funded by government, industry, state and local governments, private foundations, funds from colleges and universities, and other sources.

Industrial research is carried out by thousands of firms, large and small, although some 100 large firms account for more than 50 percent of all industrial R&D spending. The largest performers of industrial R&D are the aircraft, communications equipment, chemical, and computer and office equipment industries.

Nearly every *academic institution* conducts some research. However, about 100 universities account for more than 80 percent of all academic R&D spending.

It is estimated that there are more than 700 *federal laboratories* including FFRDCs. However, a much smaller number of these are of substantial size, with a few dozen conducting most of the R&D done in such facilities (see Box II.6).[1]

Other *nonprofit institutions* also make important contributions to national R&D performance. These include medical research institutions not associated with academic institutions, nonprofit research organizations such as Battelle Memorial Institute and Southwest Research Institute, and others.

---

[1]Federal Coordinating Council for Science, Engineering, and Technology, *Trends in the Structure of Federal Science Support* (Washington, D.C.: Office of Science and Technology Policy, 1992).

Box II.6
Types of Federal Laboratories

- Government-owned, government-operated laboratory, or *GOGO*—a laboratory owned, operated, and funded by the federal government and staffed by federal employees. Examples include NIST laboratories, NIH intramural laboratories, the National Institute of Occupational Safety and Health, and the USDA Peoria Regional Laboratory.

- Government-owned, contractor-operated laboratory, or *GOCO*—a laboratory owned and funded by the federal government and operated and staffed by a private contractor. The contractor may be a profit-making firm, a nonprofit organization, or one or more academic institutions. Examples include all of the DOE national laboratories mentioned below.

- *National Laboratory*—a large, multipurpose laboratory of the Department of Energy, including the major weapons laboratories—Los Alamos, Sandia, and Livermore—as well as Argonne, Brookhaven, Oak Ridge, Lawrence Berkeley, and others. (National Laboratories are one type of FFRDC—see next item.)

- Federally funded research and development center, or *FFRDC*—a particular form of long-term government contract with a nongovernmental organization to staff and operate a laboratory or other research center that is funded in whole or in substantial part by the federal government. Some FFRDCs are agreements to operate GOCOs, while others are contracts that support contractor-owned and contractor-staffed organizations. FFRDCs are operated by academic institutions (e.g., the Lincoln Laboratory by Massachusetts Institute of Technology) or nonprofit organizations (e.g., Project Air Force at RAND), acting alone or in consortia, as well as by profit-making firms (e.g., Sandia National Laboratories and Oak Ridge National Laboratory operated by Lockheed-Martin Corporation).

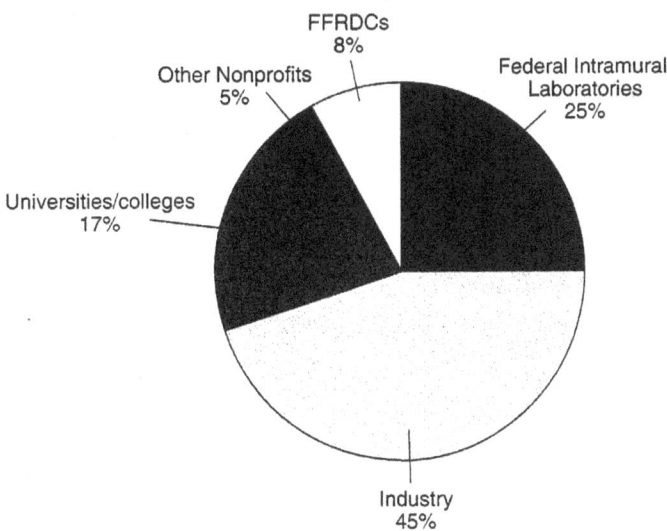

FIGURE II.6 Allocation of federal R&D funds among categories of performers, Fiscal Year 1994.

SOURCE: Data calculated from Table C-8, National Science Foundation, *Federal Funds for Research and Development: Fiscal Years 1992, 1993, and 1994* (Arlington, Va.: NSF/Division of Science Resources Studies, 1995).

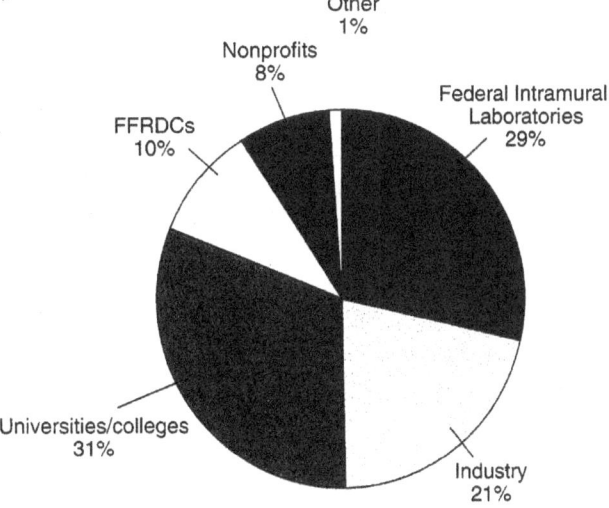

FIGURE II.7 Allocation of FS&T funds among categories of performers, Fiscal Year 1994.

SOURCE: Derived as follows: (1) R&D obligations by performer (for all federal agencies except DOD, DOE, and NASA) were taken from Table C-8, NSF, *Federal Funds for Research and Development: Fiscal Years 1993, 1994, and 1995,* forthcoming. (2) DOD, DOE, and NASA obligations for research, by performer, were taken from the same source. (3) Obligations for 6.3A by DOD were allocated among performers in the same proportions as reported in Appendix A, DOD, *DOD Response to NSTC/PRD #1, Presidential Review Directive on an Interagency Review of Federal Laboratories* (February 24, 1995). (4) Obligations for the equivalent to 6.3A by DOE in FY 1994 ($1.5 billion), as estimated by the Office of the Director of Defense Research and Engineering (see Box II.3, footnote 4), were allocated among performers in the same proportions as DOE obligations for all development in FY 1993, as reported in Table C-9, NSF, *Federal Funds for Research and Development: FY 1992, 1993, and 1994,* 1995. (5) The same approach used in 4 above was also used to allocate 6.3A-equivalent obligations by NASA in FY 1993 ($1.4 billion) among performers. (6) The funding by type of performer in 1-5 was summed and the overall percentages determined.

Supplement 3
Current Processes for Allocating Federal R&D Funds

The committee's recommendations argue for changes in how Congress and the Executive Branch allocate funds for federal science and technology. This supplement describes the current process and gives some historical background.

There Is Currently No Standard Process for Allocating Federal R&D Funds

Policymakers and the research community share control over the allocation of federal funds to R&D. In practice, decisions to allocate federal R&D funds among national goals and among federal departments and agencies are made by elected officials, senior civil servants, and congressional staff in a political process. Allocation decisions among projects and performers at the program level within departments and agencies are made by technical experts in the agencies, often with advice from the research community via formal competitive merit review or other approaches to assessing scientific and technical merit. On occasion, nongovernment scientists and engineers influence high-level strategic federal allocations to specific initiatives. Political leaders sometimes seek to influence allocations at the working level.

At all levels in the process of allocating R&D funds to various elements in the federal portfolio, there is no substitute for human judgment, informed by specialized knowledge, experience, and an understanding of the processes of research and development. There is an inherent uncertainty in anticipating the outcomes of R&D programs. Therefore, economic and financial investment models, such as cost/benefit analysis, are applicable only for those development programs for which technical and financial uncertainties are fairly well understood.

The overall federal R&D portfolio is determined in a bottom-up process. The executive and legislative branches together establish R&D budgets for departments and agencies. Historically, an "R&D budget" as such has been determined only after the fact when budget analysts learn what the overall federal R&D budget is by aggregating the results of the individual departmental and agency decisions. The Bush and Clinton administrations have sought to impose greater order on the preparation of the overall R&D budget submission, as discussed below.

Both the President and the Congress Influence the Federal R&D Portfolio

Presidents have used a variety of institutional arrangements to coordinate the formulation of R&D budgets across the departments and agencies, sometimes in hopes of orchestrating coordinated approaches to particular national problems, and other times in hopes of reducing overlap and duplication among them. Since the early 1960s, the White House Office of Science and Technology Policy and its predecessors have set up formal coordinating bodies for R&D, sometimes at the encouragement of the Congress. The Clinton administration's National Science and Technology Council is the most recent such effort, and it is too early to determine how effective it may be. However, previous bodies have had limited effect, owing to resistance by the affected agencies, the Office of Management and Budget, congressional authorizers and appropriators, and the press of political currents that are stronger than the impulse to coordinate.

There is no equivalent congressional coordinating authority for R&D (see Box II.7). The House Committee on Science, which has oversight authority over all federal nondefense R&D, comes closest, although it does not have legislative authority over the National Institutes of Health or the Departments of Agriculture, Defense, and the Interior. As it considers the President's budget, Congress and its committees frequently augment or cut proposed budgets and may replace requested R&D funds with other types of spending, with little regard for a broader interagency strategy. Even such coordinated presidential initiatives as the Global Climate Change program

---

Box II.7

CONGRESSIONAL CONSIDERATION OF THE R&D BUDGET FOR FISCAL YEAR 1996

Students of R&D budgeting have long been frustrated by the absence of a mechanism in the Congress to consider the federal R&D budget on a comprehensive basis, to address proposals from the administration for coordinated interagency R&D programs, to assess the adequacy of such funding on an aggregate basis, or to ensure against the emergence of imbalance in the federal portfolio.[1]

The 104th Congress has used some procedures that offer promise for more comprehensive congressional consideration of R&D funding in future years. In late January 1995, the House Committee on Science held a hearing on federal R&D featuring the heads of all major R&D departments and programs under its legislative jurisdiction. The House Budget Committee has established several working groups, including one on natural resources and science. One working group function, pursued with special vigor this year, was coordination with members of relevant authorization committees and appropriations subcommittees. The working group that covered science included the chair of the Science Committee (who is also vice-chair of the Budget Committee). The House Science Committee reported authorization bills within limits set by the Budget Committee in preparing its Omnibus Civilian Science Authorization bill, which also bundled together the major R&D functions under the committee's jurisdiction. The appropriations subcommittee allocations, in turn, took greater account of Science Committee and Budget Committee recommendations than in previous years. A number of important R&D budgets such as those for the National Institutes of Health and Department of Defense programs, however, do not come under the Science Committee's jurisdiction, and their R&D budgets were not handled by the same Budget Committee working group. No similar process exists in the Senate to review the R&D budget and to link different steps in the budget process across committee lines. The Senate has more committee assignments per member than the House, however, and so it is more usual for Senators to sit on multiple committees that are involved in the sequential steps of the R&D budget process.

[1] Carnegie Commission on Science, Technology, and Government, *Science, Technology, and Congress: Expert Advice and the Decisionmaking Process* (Washington, D.C.: Carnegie Commission on Science, Technology, and Government, 1991).

may emerge from the congressional budget and appropriations processes in a form quite different from that initially proposed in the budget.

International Comparisons Offer Imperfect Insight into the Desirable Level of Total U.S. National R&D Spending

Judgment, experience, and a willingness to take risks play key roles in establishing an optimal level of national R&D spending, by the federal government or by private firms, philanthropies, and other levels of government. Comparisons with R&D efforts of other leading nations offer some insights. For example, the proportion of gross domestic product (GDP) that is devoted to R&D is of some interest. In recent years, most of the larger and wealthier industrial nations have spent between 2.5 and 3 percent of GDP on R&D, including both government and private industry funding. Figure II.8 shows the percentages for Japan, Germany, and the United states through 1991.

In 1994, the Clinton administration articulated a "reasonable long-term goal" for total national R&D spending of 3 percent of GDP,[1] as compared with the present level of about 2.6 percent.[2] However, nations face different circumstances and value their national goals differently; as a consequence, they do not all spend their funds for the same purposes or in similar institutions. For example, if private industrial R&D spending is adjusted to account for the smaller role of manufacturing industries in the economy of the United States as compared with Japan or Germany, then the United States compares adequately with those nations in the ratio of R&D to GDP.[3] On the other hand, the United States has for the past 5 decades supported a large national defense R&D effort that has not existed in Germany or Japan, as well as newly emerging sectors that are research intensive but are not included in manufacturing, such as software and communications. Similarly, the United States spends a great deal more on health-related R&D than do other major nations, even when adjustments are made for the relative sizes of countries.

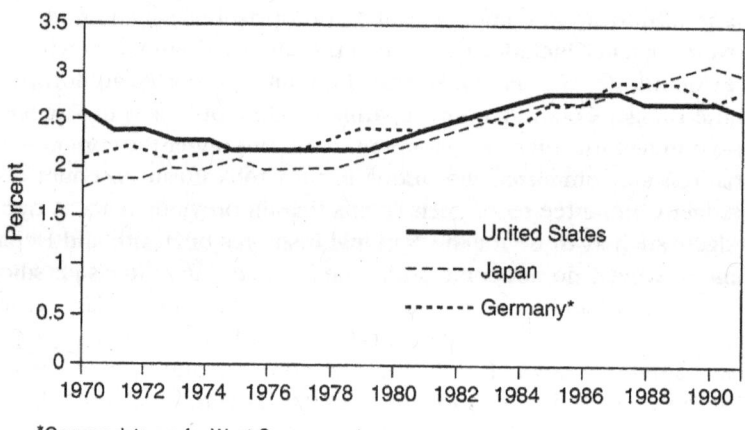

*German data are for West Germany only.

FIGURE II.8 Funding of R&D (both public and private) as a percentage of GDP for three leading nations, 1970 through 1991.

SOURCE: National Science Board, *Science and Engineering Indicators: 1993*, NSB 93-1 (Washington, D.C.: National Science Foundation), p. 375.

Strategic R&D Allocations Among U.S. National Goals Arise from a Decentralized Process

In recent years, the President's budget submission to Congress has included a section that presents the budget requests for R&D from the individual departments and agencies, as well as the total amount requested to support R&D in all of them. However, this R&D "budget" does not result from any comprehensive examination of all of the government's R&D spending. Instead, it simply presents together in one place the outcomes of the negotiations among the individual departments and agencies, the Office of Management Budget, and the President regarding their separate budget plans. The departments and agencies operate under delegations of authority from the Congress and seek to use their R&D funds to accomplish the goals set out for them by Congress. Their performance is overseen by the individual committees and subcommittees of jurisdiction. R&D programs and funding are the responsibility of numerous committees and subcommittees in both houses. Table II.2 shows the 14 committees in the House and Senate that authorize the largest R&D activities in major R&D agencies. Many of the critical congressional decisions about R&D are made in appropriations subcommittees. Figure II.9 shows the portions of the R&D budget allocated by 7 of the 13 appropriations subcommittees in each house. Figure II.10 illustrates that 14 percent of federal discretionary spending

TABLE II.2 Authorization Committees with Major R&D Programs

Department or Agency	Committee	
	House	Senate
Department of		
Agriculture	Agriculture	Agriculture, Nutrition, and Forestry
Commerce	Science Commerce	Commerce, Science, and Transportation
Defense	National Security	Armed Services
Energy		
Civilian	Science	Energy and Natural Resources
Defense	National Security	Armed Services
Health and Human Services	Commerce	Labor and Human Resources
Interior	Resources	Energy and Natural Resources
Transportation	Transportation	Commerce, Science, and Transportation
Veterans Affairs	Veterans Affairs	Veterans
Environmental Protection Agency	Science	Environment and Public Works
National Aeronautics and Space Administration	Science	Commerce, Science, and Transportation
National Science Foundation	Science	Labor and Human Resources Commerce, Science, and Transportation
Office of Science and Technology Policy	Science	Commerce, Science, and Transportation

NOTE: Main authorization jurisdictions for R&D programs are spread over seven committees each in the House and Senate. Some agencies' R&D programs are split between two or more committees because changes in Congress do not always parallel those in the Executive Branch. This table shows only main authorization jurisdictions and does not show all split authorities.

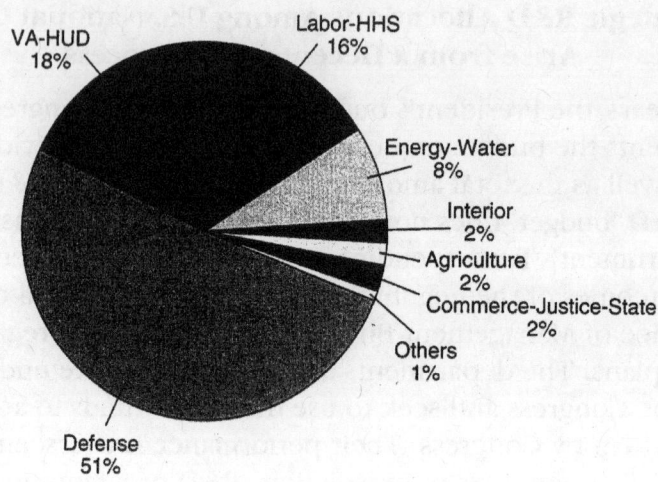

FIGURE II.9 Appropriations subcommittee roles in funding R&D.

NOTE: The $70 billion of federal R&D, as traditionally calculated, is allocated mainly by seven appropriations subcommittees each in the House and Senate. The seven subcommittees that allocate most R&D funding and the activities over which they have appropriation authority are (1) Agriculture, Rural Development, Food and Drug Administration, and Related Agencies (most USDA R&D programs; FDA); (2) Commerce, Justice, State, the Judiciary, and Related Agencies (NIST, NOAA); (3) Energy and Water Development (most DOE R&D programs; civilian aspects of DOD, such as the Army Corps of Engineers); (4) Interior and Related Agencies (U.S. Geological Survey; DOE programs on fossil fuel, coal, and conservation; and USDA Forest Service); (5) Labor, Health and Human Services, Education, and Related Agencies (NIH, Centers for Disease Control, Department of Education R&D programs); (6) National Security (most DOD R&D programs); and (7) Veterans Affairs, Housing and Urban Development, and Independent Agencies (Department of Veterans Affairs, EPA, NASA, NSF, OSTP). Each of the remaining six subcommittees allocates less than 5 percent of its appropriations authority for R&D, most far less.
SOURCE: Adapted from data provided by the R&D Budget and Policy Project, American Association for the Advancement of Science, Washington, D.C.

went to R&D in fiscal year 1995 (the figures do not take into account recisions that took effect in July 1995). The fraction of funds going to R&D is a rough measure of the trade-off between R&D and other spending within subcommittees. The higher the fraction of budget devoted to R&D, the harder it is to increase R&D without impinging on other programs and the more tempting it is to cut R&D to fund other popular programs. Funding for the National Aeronautics and Space Administration, Environmental Protection Agency, and National Science Foundation, for example, competes for dollars allocated to the same appropriations subcommittee that funds veterans' benefits and federal housing programs. Similarly, increases for the National Institutes of Health can come only at the expense of programs for education, labor, and health and human services. Given caps set by the budget process and projected steep declines in federal discretionary spending, preserving R&D funding increasingly conflicts with the desire to preserve such other programs. R&D intensiveness varies considerably, as shown for the major R&D subcommittees in Figure II.10.

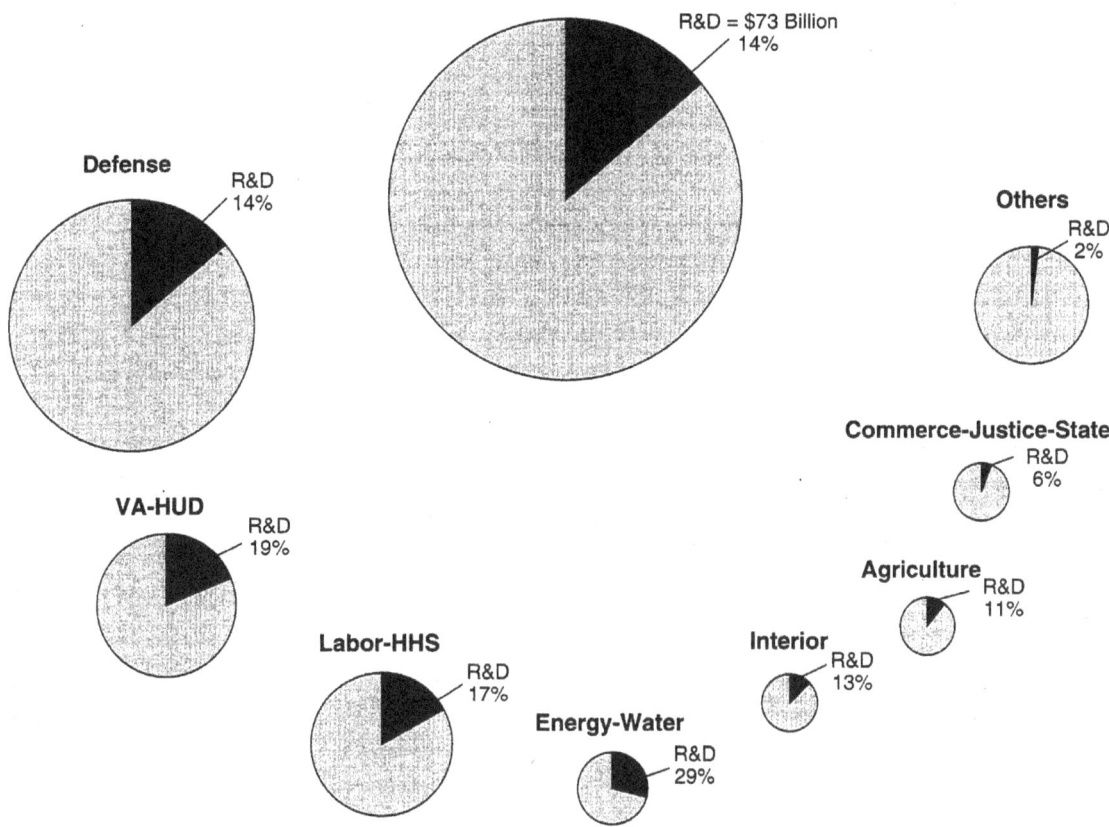

FIGURE II.10 Appropriations subcommittees' roles in funding R&D.

NOTE: Fourteen percent of the federal government's total discretionary funds went for R&D, based on fiscal year 1995 appropriations prior to the July 1995 recisions. The discretionary budget excludes mandatory federal spending and spending for entitlements, leaving $531 billion, of which $73 billion was appropriated initially for R&D by current definitions (in contrast to the committee's recommended FS&T definition discussed in Part I of this report and in Supplements 1 and 2). R&D funds are concentrated in the jurisdictions of 7 of the 13 appropriations subcommittees, which allocate from 6 to 29 percent of their discretionary funds to R&D. The largest R&D allocations are made by the Defense, Veterans Affairs-Housing and Urban Development, and Labor-Health and Human Services subcommittees. For committee names and their R&D jurisdictions, see note for Figure II.9. This information is based on 1995 appropriations prior to the July 1995 budget recisions, which reduced R&D funding by a total of $1.9 billion, taken from several programs.
SOURCE: Adapted from data provided by the R&D Budget and Policy Project, American Association for the Advancement of Science, Washington, D.C.

Tactical Allocations Among Programs, Fields, and Disciplines Are Made Largely Within Departments and Agencies, with Some Specific Congressional Direction

The mission of the National Science Foundation is to support R&D across a wide range of topics. For the National Science Foundation, tactical allocation is largely a matter of allocating funds across its various R&D programs in support of fields or disciplines.

Departments and agencies with more focused missions, however, such as the Department of Energy or the Department of Health and Human Services can, to some degree, choose whether to pursue their ultimate objectives by funding R&D or by supporting other kinds of programs in, for example, education, public health, regulation, or direct service delivery. For them, allocations to R&D result from a complex set of negotiations among the department's various bureaus, congressional oversight committees, and the Office of Management and Budget. Coordination of R&D in such agencies with that in other agencies may take a distinct second place to the intraagency struggles for resources. Most such agencies have external scientific or technical review and advisory boards, but these groups tend to focus on identifying R&D needs and opportunities and on allocating funds among projects and performers, rather than on allocations among broad objectives or between R&D and alternative implementation modalities.

Congress has always exercised its prerogatives in directing federal agencies to fund specific projects in particular locations—so called "ear-marked" activities. Not until the early 1980s, however, was this practice used for funding R&D facilities and projects. Since then, R&D earmarks have become commonplace, especially in the jurisdictions of certain appropriations subcommittees and in the budgets of certain agencies, such as the Department of Defense and Department of Energy. Earmarks to academic institutions have amounted to more than 5 percent of federal R&D funding to colleges and universities in recent years. While one rationale for such funding is that some institutions and some regions are less well prepared than others to compete for federal funds, a significant proportion of the academic earmarks has gone to institutions and states that are also successful in the open competition for federal agency funds.

Competitive Merit Review Is Most Relevant to Allocations Among Projects

One of the hallmarks of the postwar R&D system has been the detailed scientific and technical agenda influenced by the scientific and technical communities. To a first approximation, policymakers have set broad goals and directions, while members of the scientific and technical communities have designed projects, proposed priorities among them, and helped evaluate the results. Part of the "social contract" between science and government struck after World War II was that scientists would play major roles in providing advice about the scientific agenda, while policymakers would set broad strategic goals and provide the resources needed to reach them. This model has been most clearly implemented through the use of the "peer review" system to choose among research projects supported by the National Science Foundation and National Institutes of Health (see Box II.8).

The mission agencies have tended to employ their in-house scientific and technical staff to make funding decisions and to evaluate the outcomes of R&D projects focused on the government's own needs. This practice reflects the fact that government agencies must be accountable for achieving the results they set out to reach and that such work is carried out under contracts rather than grants. Increasingly, however, such agencies as the Environmental Protection Agency, U.S. Department of Agriculture, and Department of Energy have used external peer reviewers to augment the judgments of in-house staff.

Box II.8
Methods for Selecting Federal R&D Performers and Projects

A number of approaches are used to decide which R&D projects receive federal funds, how much should be spent, and who should conduct the work. The approach used depends on the nature of the work, its relationship to specific government missions, and the history and culture of different research communities, programs, and agencies.

Traditionally, agencies such as the National Science Foundation and National Institutes of Health that make grants to universities to support fundamental scientific and engineering research have used some form of prospective *peer review* to judge the quality of competitively submitted project proposals. Peers are established working scientists or engineers from diverse research institutions who are deeply knowledgeable about the field of study and who provide disinterested technical judgments as to the competence of the researchers, the scientific significance of the proposed work, the soundness of the research plan, and the likelihood of success. Since the early 1980s, NSF has asked peers also to take into account the utility of the proposed research to the nation and its potential for contributing to graduate education and to the infrastructure of science itself. Since the middle 1980s, NSF has used the term *merit review* to indicate both that proposals are judged on their merits and that NSF program officers also have the authority to take into account various general policies of the Foundation when making awards. NIH makes limited use of a second level of review by institute councils that take into account national relevance and direction. Some programs in other departments and agencies, including the Department of Energy, National Aeronautics and Space Administration, U.S. Department of Agriculture, and Department of Defense, employ variants of peer or merit review. The various departments and agencies differ in the degree to which their program managers are bound to follow the recommendations of peer and merit reviewers in making awards. Practices vary even within the NSF and NIH across research fields and areas.

Other agencies, including the Office of Naval Research and Advanced Research Projects Agency, use a *strong program manager* approach to prospective assessment of the scientific or technical merit of research proposals, particularly those that are of a more fundamental nature. Strong technical staff members have responsibility for being well informed about the state of the art of their specialties and for identifying and recruiting investigators to conduct research that they deem to be of greatest importance to the agency's mission. Program managers often devote considerable energy to soliciting the views of peers about these matters but usually are not bound to heed their advice.

Agencies seeking to contract for performance of R&D projects of direct interest to the government in industrial or other nongovernment organizations typically conduct competitive procurements for R&D services, using *government technical employees* and, occasionally, consultants to judge the prospective merit of contract proposals. This approach has much in common with standard procedures used by the federal government to procure other goods and services.

Federal laboratories use several approaches to project selection. In most cases, however, *on-site technical and unit managers share responsibility with agency program managers* for selecting project topics and performers. In some cases, the external peer community is asked for advice on specific projects, and in other cases, on an entire program of activity. Sometimes such advice is obtained on a prospective basis; sometimes it is obtained via formal reviews of ongoing or completed research activities. In some agencies and some programs, proposals to begin new projects at the federal laboratories compete across several laboratories or even with proposals submitted from academia or industry.

Formula funding is used by a few programs, principally in the USDA, to allocate R&D funds among performing institutions such as the land-grant colleges and universities.

Executive agency decisions about R&D allocations to institutions and projects have increasingly been specified in detail by *congressional appropriations committees*. These allocations often do not reflect the considered judgments of scientific experts or the funding agencies, and they often are determined instead by individual members of Congress acting on behalf of constituents.

Supplement 4
Interactions Between Federal and Industrial Funding and the Relationship Between Basic and Applied Research

Continuing innovation is the only way to foster long-term economic growth without discovering entirely new resources. Advances in science and technology are essential to innovation, although innovation also involves many additional factors. In the last half century, the federal government's role has almost always been crucial, and often dominant. The nation has become more dependent on science and technology, and sustaining a robust capacity for research and development is more important than ever. Astronauts have walked the face of the moon and returned, and astrophysicists have probed the origins of the universe. The physical sciences have also been the source of innumerable inventions—lasers, microelectronic devices, and fiber-optic networks, to name just a few—that have in turn enabled practical applications such as satellite communications, computers, and gains in productivity throughout the economy. Past revolutionary advances in biology—unraveling the double helical structure of DNA in 1953, discovering recombinant DNA technology in the 1970s—and today's exploding molecular genetics and integrative biology have just begun to illuminate the immense complexity of life. These fundamentally important discoveries also are linked to the capability to design new drugs and diagnostic technologies in medicine, new approaches to problems in agriculture, and technologies for environmental improvement.

The dramatic increase in life expectancy during this century is one indicator of scientific discovery and technical progress. Figure II.11 shows that in 1900 life expectancy at birth, even for the richest people, was only age 55, yet for all but the world's poorest today it is over 70. Every year during this century, approximately 2 months have been added to life expectancy. The change has been gradual, almost unnoticed in daily life, but fundamentally important. Sanitation, nutrition, transportation, communication, and other technologies have combined with biomedical research and medical technologies to produce this profound demographic shift.

In the 5 decades following World War II, the U.S. federal government steadily increased its support for science and technology. As a result, the United States moved into a position of preeminence in virtually all areas. We became the leaders in high-technology industries such as aircraft, chemicals, computers, software, pharmaceuticals, and biotechnology. We developed the most effective system in the world for creating new technology-based businesses.

Our National System of Innovation Depends on Complicated Interactions Between the Public and Private Sectors

A complex set of institutions and actors contribute to the strength of the U.S. science and technology base. The examples of important discoveries in medicine and in computing and communications technologies depicted in Figures II.12 and

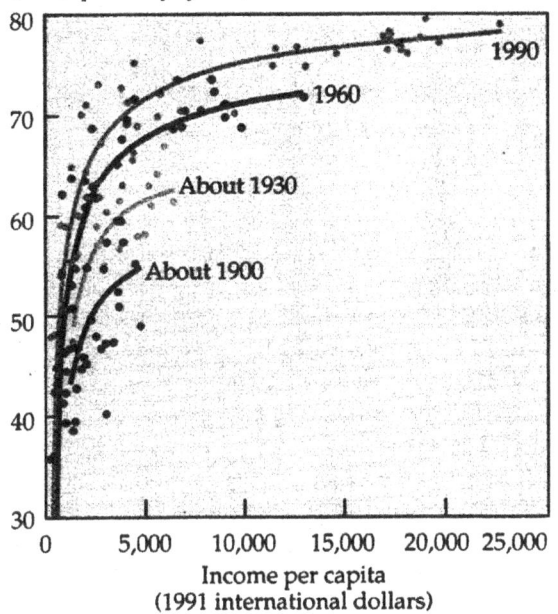

FIGURE II.11 Life expectancy and income per capita for selected countries and periods.

NOTE: "International dollars are derived from national currencies not by use of exchange rates but by assessment of purchasing power. The effect is to raise the relative incomes of poorer countries, often substantially."
SOURCE: Samuel H. Preston, Nathan Keyfitz, and Robert Schoen, *Causes of Death: Life Tables for National Populations* (New York: Seminar Press, 1972), as reprinted in The World Bank, *The World Development Report 1993: Investing in Health* (New York: Oxford University Press, 1993).

II.13 illustrate how the different parts of our system interact to produce changes that ultimately lead to increases in the quality of our lives.

Consider, for example, Figure II.12, which illustrates the steps leading to reductions in mortality from high blood pressure.[1] From the 1930s to the 1960s, research funded by private nonprofit groups, the Veterans Administration, and the National Institutes of Health revealed dietary and behavioral risk factors associated with high blood pressure. An early and important step was finding a way to measure blood pressure quickly and cheaply, and to correlate those measures with diseases. Rigorous epidemiological studies confirmed suspected links between high blood pressure, stroke (and later, heart disease), and premature death. Parallel clinical trials demonstrated that treatment for lowering blood pressure prevented stroke, death from heart disease, and cardiac and renal failure. The National High Blood Pressure Education Program, built on these findings, commenced in 1972. Since then, changing social norms, individual exercise and diet decisions, and better medical management have reduced the incidence of hypertension by more than a third, and reduced stroke mortality by over 60 percent,[2] a remarkable achievement. For millions of Americans, a broad base of research—spanning the full range from social and behavioral research to molecular biotechnology—has meant the difference between life and death.

As epidemiological and behavioral research progressed, a complex web of biological factors also was uncovered through clinical investigations and basic biological research. This line of research was funded predominantly by the federal government, and supplemented by hospitals and private sources. Private pharmaceutical firms made investments comparable in magnitude to federal funding, but focused on narrowing the search for specific agents and clinical testing to prove their worth. Drugs lower blood pressure by reducing fluid retention (diuretics), by influencing nerve impulses transmitted to the heart and blood vessels (beta-blocking agents), and by reducing resistance to blood flow in small peripheral arteries (cal-

Timeline of Hypertension Research and Drug Development

Diet, Exercise, Smoking, Behavior
- National Heart Institute Established (1948)
- Framingham Study (1950–)
- Surgeon General's Report on Smoking (1964)
- Johnson Administration War on Cardiovascular Disease (1960s)
- National High Blood Pressure Education Program (1972–)

Diuretics
- Clinical Research (1950s–)
- Thiazides (MERCK, CIBA, then WYETH-AYERST, ABBOTT, RORER, LEDERLE) — Loop Diuretics (HOECHST, ROCHE, MERCK)
- Spironolactone (SEARLE)
- Triamterene (SMITH-KLINE)
- Amiloride (MERCK)

β-blockers
- Alquist Postulates Receptor Types (1948)
- Propranolol (ICI)
- Others (MERCK, ABBOTT, SEARLE, GEIGY) (20,000,000 Prescriptions)

Angiotensin Converting Enzyme Inhibitors (ACE Inhibitors)
- Enzyme Action of Renin Discovered 1940
- Angiotensin Types
- Snake Venom Blocks Renin Formation
- Captopril (SQUIBB)
- Enalapril, Others (MERCK, BRISTOL-MYERS, CIBA, STUART)

Calcium Channel Blockers
- Calcium and Muscle Contraction 1883–1913
- Fleckstein and Kaufman Show Calcium Antagonism
- Verapamil (HOECHST, KNOLL, then LEDERLE, SEARLE)
- Clinical Use for Hypertension
- Nifedipine (MILES, PFIZER)
- Diltiazem (DOW)

Timeline axis: 1946, 1950, 1960, 1970, 1980, 1990

Legend:
- ■ Federal or private nonprofit-funded R&D
- ▨ Industrially funded R&D
- ■ $1 billion sales (or 20,000,000 prescriptions, for β-blockers)

FIGURE II.12 Steps in discovering how to prevent and manage high blood pressure.

NOTE: Dramatically reduced mortality and disability from stroke and heart disease have followed from better prevention, identification, and treatment of high blood pressure. The top section shows federal programs that laid the groundwork for clinical treatments indicated lower in the diagram. Drug strategies now supplement diet, exercise, and regular medical monitoring as mainstays of medical management. Several different types of drugs are used. The four most common classes are diuretics, beta-blockers, angiotensin converting enzyme inhibitors, and calcium channel blockers, whose developmental history is summarized in the figure. Drug classes are noted along the left margin, and specific agents are noted in the diagram, with the corresponding companies in parentheses. Drug development has been based on a mix of federal and privately funded R&D. The federal government has supported basic biological research, epidemiology, behavioral and social science, and clinical research for decades. Private firms have developed new drugs to reduce blood pressure once it is detected. Often, federally funded research has preceded private R&D, but in several cases, private firms have discovered drugs that were only later found to be useful in lowering blood pressure. Clinical research has been necessary in many ways at many stages, supported by a mix of funds derived from patient services, federal programs, and private-firm investments.

SOURCE: Rebecca Henderson, "The Evolution of Integrative Competence: Innovation in Cardiovascular Drug Discovery," *Industrial and Corporate Change* (No. 3, Winter): 607-630, 1994; the historical research of Harriet Dustan (University of Vermont), Edward Roccella (National Heart, Lung, and Blood Institute), and Howard Garrison (Federation of American Societies for Experimental Biology); National Heart, Lung, and Blood Institute, *National High Blood Pressure Education Program: 20 Years of Achievement* (Bethesda, Md.: National Institutes of Health, 1992); and Thomas P. Gross, Robert P. Wise, and Deanne E. Knapp, "Antihypertensive Drug Use: Trends in the United States from 1973 to 1985," *Hypertension* 13 (Supplement 1):I-113–I-118, 1989.

cium antagonists and angiotensin converting enzyme, or ACE, inhibitors). The use of calcium channel blockers came from clinical tests of various compounds that were first made by pharmaceutical firms. Knowledge of how these compounds worked and how they could best be used came years later, mainly through federally funded research. In contrast, the ACE inhibitors were developed by drug companies through a logical progression of discoveries that built on decades of publicly funded research. Private investment was essential, but federal investment was equally important at many stages, both leading and following privately funded research. Almost all the important technical decisions, in both public and private sectors, were made by those educated in research universities and trained at least in part through federally funded research.

The story in information technologies involves different agencies and domains of science, but the lessons are similar.[3] Lynn Conway of Xerox and Carver Mead of the California Institute of Technology in the 1970s conceived of "silicon foundries," where graduate students, their professors, and others could have computer chip designs fabricated into integrated circuits. Their idea won federal support and became the heart of the very large scale integrated (VLSI) circuit program supported by the Advanced Research Projects Agency in the Defense of Defense. NSF joined the program, broadening access to VLSI fabrication services—the foundries. On a parallel track, the network that later became the Internet (first as ARPANet) was used to send designs to the foundries, which then created and shipped the chips, reducing cost and increasing speed. What once took months now took days. The impediments to chip design diminished; graduate students felt free to experiment and innovate; even radical designs for chips became practical.

The foundries and other components of ARPA's VLSI program had spectacular results: a renaissance in computer design, universities creating VLSI programs, the beginnings of three-dimensional graphics, and initial efforts in reduced instruction set computing (RISC), now in use in millions of computers. RISC computing originated at IBM but was adopted only after a period of federally funded research that made its applications readily apparent, at which point several firms in addition to IBM invested in it. Several major corporations grew directly out of the VLSI program.

Decades of federal and industrial investments in information technology led to the creation of the elements—from three-dimensional graphics to windows to local networks—now embedded in the way we work, obtain and share information, and teach our children. The dynamic interactions between federally funded academic R&D and industrial R&D made the United States dominant in information technology, which strengthened the nation's competitiveness and also provided advantages in other sectors throughout the economy that depend on information technologies, such as finance, entertainment, communications, education, and transportation (see Figure II.13).

As has been detailed in the case of information technology and is evident also in medicine and in many other fields highly dependent on science, the history of innovative development with significant social and economic benefits points to several major conclusions:[4] (1) research has consistently generated large payoffs; (2) these payoffs often take years or decades to be realized; (3) while the time from discovery to market may be long, the transition from science to technology is more

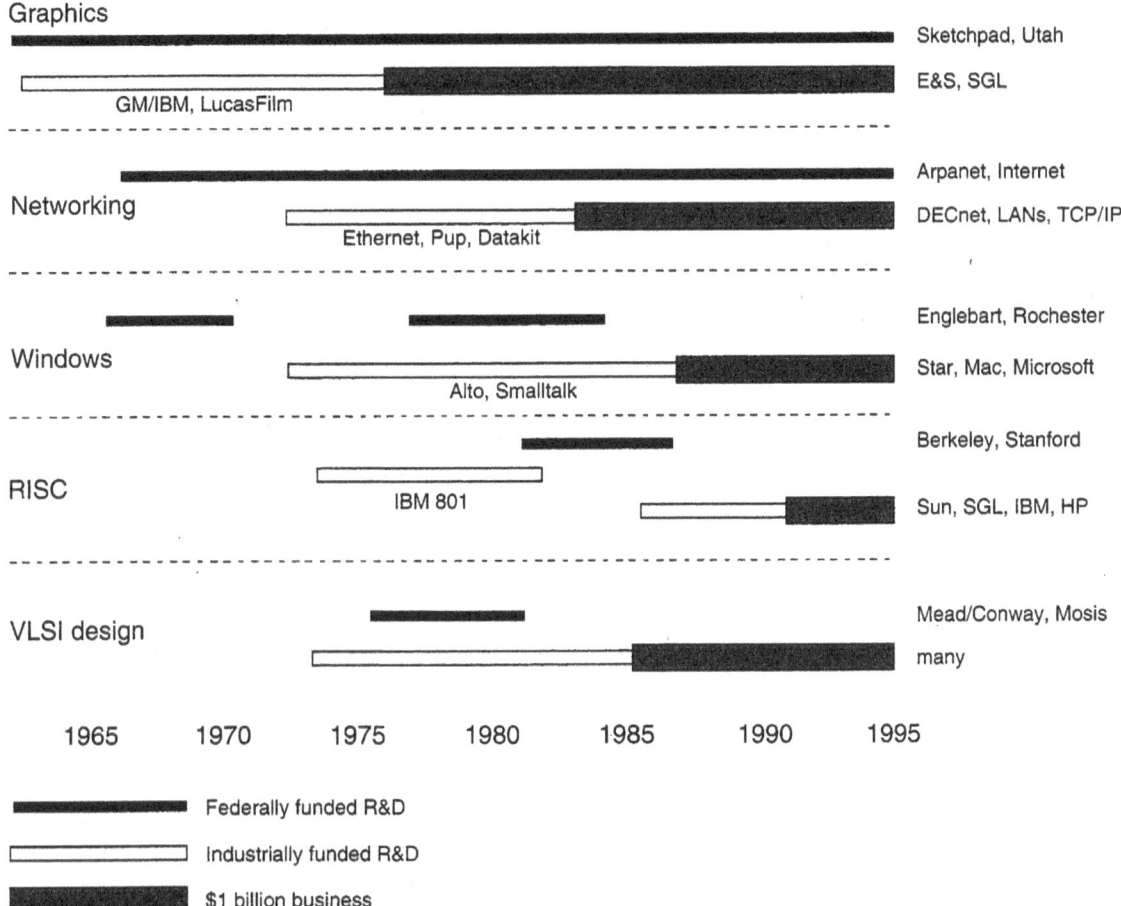

FIGURE II.13 Technological developments in computing.

NOTE: The productive and profitable interactions between federally and privately funded R&D are apparent in this time line of the development of several important computer technologies. These include computer graphics; networks; use of icons, buttons, and other "user-friendly" methods now commonly known as "windows"; reduced instruction set computing (RISC), which simplifies and speeds computer operations; and very large scale integrated (VLSI) circuit design, which has proved crucial to many manufacturing and design improvements. The institutions at which federally funded work was begun are noted along the right margin, as are the companies that developed and eventually commercialized the technologies. In many cases, the federally funded work was conducted at universities, but some was done in industry. Note that privately funded R&D preceded federal R&D in the cases of VLSI and RISC, and yet federal funding was nonetheless crucial in enabling the creation of ideas realized ultimately in commercial applications.
SOURCE: Adapted from Figure ES.1 in a report by the Computer Science and Telecommunications Board, National Research Council, *Evolving the High Performance Computing and Communications Initiative to Support the Nation's Information Infrastructure* (Washington, D.C.: National Academy Press, 1995), p. 2.

sudden; (4) unexpected results are often the most important (e.g., electronic mail and computer "windows" software methods were not the intended products of research programs that spawned them; many drugs used for hypertension were first developed for other purposes); (5) research stimulates communication and interaction, with complex interactions between industry and academia; (6) research trains people who start new companies, join established firms, and enter crucial positions in industry and government where their technical background enables better management decisions; and (7) research entails risks, so that some objectives are not reached, but new ones—often more important ones—replace them.

Practical applications are often impossible to predict from any one scientific discovery that is nonetheless crucial to the ultimate outcome, and the best path to the desired use must adjust continually to surprising sources of new knowledge. Norman Ramsey's Nobel Prize-winning work in physics was seminal in the development of atomic clocks that enabled the global positioning system (see Box II.2), magnetic resonance used for medical imaging, and synchrotron radiation used in the manufacture of integrated circuit chips. Yet none of these immensely practical benefits was evident when he did his research. He remarked upon receiving the 1994 Vannevar Bush Award, "I would have had difficulty in justifying most of my research on the basis of future applications either I or anyone else would have foreseen."[5]

The government role in supporting the federal science and technology (FS&T) base is crucial in almost all the technologies. In some cases and at some stages, it is the dominant factor. The critical period for federal investment is often, but not always, at the beginning. Federal support for basic science is often necessary, but federal support for applied research and fundamental technology development is also essential. Some new technologies do build logically on scientific discovery arising from federally funded basic science, but private research and development often turn up items that pose questions for science or require a period of government-supported inquiry before they become appropriate for further development in the private sector. Federal support often comes from different agencies, at different times, and for different reasons.

Research and development, and the ensuing innovation system of which they are essential components, depend not only on the basic science supported by the National Science Foundation, but also on mission-oriented research and development of the National Institutes of Health, Department of Defense, National Aeronautics and Space Administration, Department of Energy, Environmental Protection Agency, U.S. Department of Agriculture, Department of Commerce, and other agencies and departments.

The Distinction Between Basic and Applied Science Is Often Difficult to Make and Is Rarely Decisive in Defining the Federal Role

Historically, the federal government has provided funding for a variety of long-term, high-risk research and technology development programs. In some cases this support is motivated by the need to solve specific problems such as developing a new aircraft, breaking a code, or finding a way to treat specific diseases. The resulting activity conventionally is described as applied research. In other cases, government support is provided for pure science. Some projects are clearly applied. Others are clearly basic. Basic research usually is supported in the expectation that it ultimately will link to practical use; applied research usually is intended to address a specific problem, although it can spawn new fundamental inquiry.[6]

Some discussions of the differences between basic and applied research suggest that the process must start with basic research in universities, which produces new ideas. In this view, private firms apply discoveries to practical problems and use them to develop commercial products. Sometimes the discovery process works this way, but often it does not. The flow of people, knowledge, and "know-how" between publicly and privately funded research organizations goes both ways, with different net flows at different times. The typical patterns differ among industrial sectors and scientific disciplines; there is no one template for innovation. For every case like that of information technology—where academic research in computer science and engineering led to the creation of many new firms—one can point to a counterexample like digital electronics, where the development of the transistor in the private sector caused an expansion of solid-state physics in universities. Even when a clear distinction between basic and applied research can be made, therefore, it is often not useful in guiding choices about whether it is a proper subject for federal support.

A more severe problem is that most federally funded research is at once both applied and basic. In the standard definition, basic research is the pursuit of knowledge without thought of practical application. The first part is true—that science is intended to produce new discoveries—but the implication that this necessarily entails a sharp separation from thoughts of usefulness is just plain wrong. Sometimes it is true, but far more often it is not, especially in science supported by mission-oriented agencies. Basic optics is one of the oldest fields in physics. Thirty or forty years ago, it was hard to see what applications it might have beyond lens design for cameras and telescopes. With the unexpected discovery of the laser and its application in fiber-optic communications, optics has turned out to be immensely practical, and is essential to modern telecommunications networks. Louis Pasteur's career was replete with contributions to basic biology as well as innovations in medicine, beer brewing, wine making, and agriculture. Organic chemistry and analytical chemistry have always been coupled to pharmaceuticals, specialty chemicals, and other industrial interests. Basic materials science bears on electronics, instrumentation, aeronautics, and many domains of manufacturing. Gregor Mendel was studying how to improve crops when he discovered the basic laws of genetics, and characterizing DNA's double helical structure in 1953 led 2 decades later to practical applications through recombinant DNA technology, with impacts not only on biomedical research but also on pharmaceutical manufacturing, agriculture, and environmental remediation. The practical uses of applied research are generally more obvious and direct, but basic research also can have foreseeable practical aims.

"There are two kinds of research—applied research and not-yet-applied research." Nobel laureate Lord Porter, former president of the Royal Society.[7]

The federal responsibility for basic research is accepted widely. The large social benefits that can come from federal support for specific kinds of applied problem solving and exploratory development are not as well recognized. Historically, a large fraction of federally funded research has been directed at applied problem solving and fundamental technology. For example, when the Department of Defense funded the creation of computer science as a new academic field, it accurately anticipated real national security needs. Government support for the development of new problem-solving tools and technically trained people differs from the support for basic physics provided by the National Science Foundation, but it nonetheless has profound effects on fulfilling national needs, sustaining our economy, improving our way of life, and contributing to all areas of scientific investigation.

Government Has Traditionally Supported Enabling Technology and Education

There is no reason to abandon the historical balance between support for science on the one hand and enabling technology on the other. Industrial funding builds not only on basic research, but also on federally funded R&D aimed at government functions. The productivity of industrial R&D depends on a balanced federal R&D portfolio that spans a broad range of applications. A strategy that focuses federal support unduly on basic science risks losing the benefits of applied research supported by mission agencies, which historically have been important in generating public benefits. That is one reason that the committee did not distinguish between basic and applied research when defining the FS&T budget.

In the division of labor between the public sector and the private sector, the private sector ultimately will be responsible for the final stages of commercial application and product development. On this there is no disagreement. Because of its efforts in these areas, the private sector will provide more support for applied research and technology development than the federal government does now or could at any time in the foreseeable future. But there can be confusion about the federal role in supporting applied research versus its funding of commercial technology development in industry, whether through individual firms or in consortia. This is an area of active controversy that the committee addresses in Part I of this report. It is important to point out here that the debate about federal funding, or subsidies, to industry is conceptually different from that about federal support for basic versus applied research for public missions, to foster enabling technologies, and to educate leaders in science and engineering.

A distinction between "basic" and "applied" generally is not useful as the decisive criterion that defines a proper federal role, except when the application area is an existing commercial market where industrial applied research usually will predominate. Federal leadership is indeed essential for basic research, because industry does not support it except in a limited way and under unusual circumstances such as near-monopoly positions that are now rapidly disappearing.[8] Five decades of history make clear that the federal government is positioned uniquely to support the training of people and the development of new technologies that are not specific to a particular product or service. Private firms, responding to forces that operate through the market, will determine what specific products and services result and will support their final development and commercialization.

The government also must maintain a core of applied scientists whose work serves as a bridge between the problem-solving efforts of private firms and the research efforts of basic scientists. Government can and should sustain those areas of science and technology that support inherent government missions, such as national defense, technical standards setting, regulation, or public health. In these areas, publicly funded scientists and engineers can take a discovery—such as a new class of high-temperature superconductors—out of the laboratory of a private firm and move it quickly onto the agenda for inquiry in basic physics. Or they can take a new technology and use it—applying it to nuclear waste cleanup, developing vaccines for U.S. troops headed abroad, or defining the exact length of a meter or a second with the greatest precision available at the time.

Federal funding for science and technology development also helps educate and train not only those scientists and engineers who continue on to perform research and development in both the public and private sectors, but also those whose work involves making technically informed management decisions about corporate strategy and finance. The history of technological advance throughout this century points to an abiding truth: "The primary function of universities is to give students the intellectual underpinnings to contribute as professionals in our society."[9]

Federal Support for Basic Research Continues to Be Essential

Just as government support for applied science and technology development remains a wise investment, so also is continuing investment in basic science essential to future innovation and progress.[10] Innovation now occurs too rapidly for one player to wait until another's job is done. Research and development are not separate, serial activities, but parallel and interdependent. New knowledge is most useful to people and institutions that see it first and can exploit it quickly, and that have ready access to those who discover it. The ability to identify technological opportunities emerging from research is now a principal factor determining success in many industrial sectors. The increased importance of science in high-growth areas of the world economy puts a premium on strong linkages between science and technology, and makes innovation far more difficult without a strong indigenous science base. This circumstance underscores the importance of federal support for the science and technology base as the main source of "patient capital" that builds knowledge and supports all firms. Continuing federal support for basic research is the foremost recommendation of those in industry itself.[11]

Today, the product cycle is contracting in high-technology sectors throughout the world. Software applications may be replaced after a year or two, and a computer model every three or four years. Private firms are driven by short-term market needs and demands for quick returns on science and technology investments. They must focus on improving existing products. Communications and computing were once the province of monopolies and near-monopolies that no longer exist because of federal policy and international competition. With a few exceptions, such as pharmaceuticals where patent protection is strong, support for science and technology that will not return benefits quickly is becoming more difficult to justify in the private sector, because stockholders cannot see the immediate benefits of R&D expenditures.

While the time from discovery to market has not shortened nearly as much in pharmaceuticals as in software or other sectors, drugs now are replaced more quickly once they enter the market because new agents are discovered that have stronger action or fewer side effects, and generic drugs are introduced quickly after a patent expires. Pharmaceutical firms have concluded that survival depends on increasing the pace of innovation, introducing more products in less time, and data show that strong connections to basic research performed outside the firm, as well as strong R&D capacity within it, predict success in discovering new drugs.[12] Pharmaceutical executives report that their products depend more on federally funded science than any other industrial sector, and patent statistics bear this out.[13] Thus even in a sector where private firms' R&D investments are high, and encompass some basic research, the federal role remains vital.

In the 1970s and 1980s attention turned to the dramatic technological advances made in Japan. Success there depended on improving technologies discovered elsewhere more than on Japanese science. The Japanese postwar strategy followed the "technology first" strategy pursued with equal success by the United States early in this century. In light of Japan's economic success and U.S. history, some observers began to question why U.S. taxpayer dollars should support basic research at all.

The case histories tracing drug discovery and advances in computing and communications show that it can still take decades before the practical uses of knowledge arising from disparate fields become apparent. But once commercial opportunities are apparent, it is a flat-out race from the laboratory to the market. A "technology first" strategy falters as the time scale from discovery to application shortens, as the stock of untapped but freely available existing knowledge is depleted, and as many nations attain technological expertise. As one analysis of links between patents and citations to scientific literature noted, "The areas which are leading the industrial growth of the West are just those areas that are very science intensive, and it is hard to imagine sustained industrial growth in any country without a strong competence in the scientific fields which so closely underlie these modern technologies."[14] Successful nations must not only build and sustain a firm technological base, but must also in the future make new discoveries and translate them into new technologies. Such achievements require a broad and deep base of science and technology, comprising not only those performing it but also those who monitor and use it. Those with foresight, even in Japan which now lacks a substantial science base, have recognized that neglect of science is a potentially fatal weakness in life on the technological frontier.[15]

"Until now Japan has depended primarily on foreign nations for the creative activities that generate the knowledge and technology for innovative products.... [F]rom now on Japan will have to create, ahead of other nations, knowledge and technology that will lead to new products and markets."[16]

Today, Japan, several European countries, and many emerging nations can take advantage of new discoveries, a position the United States occupied alone several decades ago. Other nations have built technological capacities that rival, and in some areas surpass, those in the United States. They have strong education systems and pursue national policies to foster innovation. But none can match the breadth and depth of U.S. science and the fluidity with which results and people move back and forth between the university and the private sector. Federal science and technology, and its connections to a robust private sector, are among this nation's most important comparative advantages.

Government Support for Scientific and Technical Public Goods Is Central to Creating National Economic Advantage

Federal funding—for basic research, applied research relevant to government missions, development of technology, and education and training in universities—encourages new firms to enter high-technology areas. Applied academic research funded by the federal government has helped produce many small high-technology firms. Students and professors move from the university to existing small firms. Sometimes they start new firms. Often they join well-established firms and rise through the ranks to make critically important decisions. New firms may grow into industrial giants or be swallowed by larger firms that incorporate their technologies. Patent rights for new discoveries derived from federally funded research go to the research institutions, giving them financial incentives for commercial application. Sun Microsystems and Silicon Graphics among computing companies, and Amgen and Genentech in biotechnology, did not exist 15 years ago. All were started from a base of academic science. Today they are major firms in their respective industries. These and other successes well up from the science and technology base supported by the federal government, which fosters competition and helps introduce new firms that champion emerging technologies.

A fear that the benefits from federal support for university research will flow immediately to foreigners is misplaced. History suggests instead that where research takes place has a direct effect on where it is put to use. The high-technology firms clustered along Route 128 in Massachusetts, in the Silicon Valley in California, in suburban Maryland, and in Austin, Texas, all congregated around major federally supported university or government research centers. If industrial use and centers of capital were the decisive factors, the foremost centers for biotechnology and computer firms should have located instead near Tokyo, Frankfurt, Paris, London, or New York City.

Endnotes

Supplement 1

1. National Science Board, *Science and Engineering Indicators, 1993*, NSB 93-1 (Washington, D.C.: National Science Foundation, 1993), pp. 328, 337.
2. Vannevar Bush, *Science—The Endless Frontier*, Appendix 3, "Report of the Committee on Science and the Public Welfare" (Washington, D.C.: U.S. Government Printing Office, 1945).
3. Bush, *Science—The Endless Frontier*, 1945.
4. See, for example, Bruce L.R. Smith, *American Science Policy Since World War II* (Washington, D.C.: The Brookings Institution, 1990); Jeffrey K. Stine, *A History of Science Policy in the United States, 1940-1985*, Science Policy Background Report No. 1 prepared for the Task Force on Science Policy, Committee on Science and Technology, U.S. House of Representatives (Washington, D.C.: U.S. Government Printing Office, September 1986).
5. Bush, *Science—The Endless Frontier*, 1945.
6. The Dana Alliance for Brain Initiatives, *Delivering Results: A Progress Report on Brain Research* (New York: The Charles A. Dana Foundation, 1995); Sandra Ackerman for the Institute of Medicine, *Discovering the Brain* (Washington, D.C.: National Academy Press, 1992); Office of Science and Technology Policy, *Maximizing Human Potential: Decade of the Brain 1990-2000* (Washington, D.C.: Federal Coordinating Council for Science, Engineering, and Technology, Executive Office of the President, 1991); Constance Pechura and Joseph B. Martin, eds., *Mapping the Brain and Its Functions* (Washington, D.C.: National Academy Press, 1991).

Supplement 2

1. Calculated from Table C-8 in National Science Foundation, *Federal Funds for Research and Development: Fiscal Years 1993, 1994, and 1995*, NSF 95-334 (Arlington, Va.: NSF/Division of Science Resources Studies, forthcoming).
2. National Science Foundation/Division of Science Resources Studies, "Data Brief: U.S. R&D Funding Will Not Pick Up in '95," No. 13 (October 18, 1995), Appendix Table B-3. The sample design for estimating industry expenditures was revised for 1991 and later years. The effect of the changes in sample design was to increase the estimate of industry R&D and thus reduce the federal share of the national R&D total by several percentage points compared with earlier surveys. Industry has contributed about 59 percent of national R&D investment in recent years. Moreover, industrial support for R&D has increased over the past 2 decades, in most years more rapidly than federal funding.

    A recent Battelle Memorial Institute survey also estimated that the federal government is funding 36 percent of the national investment in research and development in 1995. See "Funding Forecast," *R&D Magazine*, January 1995, pp. 4LS-7LS.
3. National Science Foundation, "Data Brief: U.S. R&D Funding Will Not Pick Up in '95," 1995, Appendix Table B-2. The changes in sample design in the survey of R&D expenditures by industry for 1991 and after have the effect of reducing the federal share of R&D funding of industry R&D compared with earlier surveys by NSF.
4. All constant-dollar R&D and FS&T data are in Fiscal Year 1987 dollars, calculated from current-dollar data using the GDP deflators following standard NSF and OMB practice. Fiscal Year 1995 estimates have not been adjusted to account for recisions totaling nearly $2 billion in R&D budget authority enacted by Public Laws 104-6 (April 1995) and 104-19 (July 1995) (see American Association for the Advancement of Science, *Interim Report on Congressional Appropriations for R&D in FY 1996* (Washington, D.C.: AAAS, 1995), Table A, for more details on recisions by agency and program).
5. American Association for the Advancement of Science, *Research and Development FY 1996, AAAS Report* (Washington, D.C.: AAAS, 1995), p. 3; *Interim Report on Congressional Appropriations for R&D in FY 1996*, 1995, p. 6.

Supplement 3

1. Office of Science and Technology Policy, *Science in the National Interest* (Washington, D.C.: Executive Office of the President, August 1994), p. 15.

2. National Science Foundation (including data from Department of Commerce), data for the United States, Table B-15, "National Expenditures for R&D as a Percentage of Gross Domestic Product, by Source of Funds: 1953-94," *National Patterns of R&D Resources: 1994*, NSF 95-304 (Washington, D.C.: National Science Foundation, 1995), p. 71; data on other countries, Table B-20, "National Expenditures for the Performance of R&D as a Percentage of GDP, by Country: 1970-93," p. 77.

3. See *Forging the Future: Policy for American Manufacturing*, report of the Manufacturing Subcouncil to the Competitiveness Policy Council (Washington, D.C.: Competitiveness Policy Council, March 1993), pp. 218-219.

Supplement 4

1. Material about hypertension treatments is based largely on research undertaken by Rebecca Henderson and her colleagues (Sloan School of Management, Massachusetts Institute of Technology) through a project titled "Understanding the Role of the Public Sector in Pharmaceutical Innovation," and on the historical research of Harriet Dustan (University of Vermont), Edward Roccella (National Heart, Lung, and Blood Institute), and Howard Garrison (Federation of American Societies for Experimental Biology).

2. Historical research of Harriet Dustan (University of Vermont), Edward Roccella (National Heart, Lung, and Blood Institute), and Howard Garrison (Federation of American Societies for Experimental Biology).

3. Illustrations from telecommunications and computing, along with many conclusions in this section, are taken from a report of the Computer Science and Telecommunications Board, National Research Council, *Evolving the High Performance Computing and Communications Initiative to Support the Nation's Information Infrastructure* (Washington, D.C.: National Academy Press, 1995).

4. Computer Science and Telecommunications Board, National Research Council, *Evolving the High Performance Computing and Communications Initiative to Support the Nation's Information Infrastructure*, 1995.

5. Norman F. Ramsey, Lyman Physics Laboratory, Harvard University, "Response to Vannevar Bush Award," personal communication to Robert Cook-Deegan, National Academy of Sciences, June 19, 1995.

6. Many of the points in this and the next section are adapted from Donald E. Stokes's forthcoming book, *Pasteur's Quadrant: Basic Science and Technological Innovation* (Washington, D.C.: Brookings Institution, 1996).

7. George Porter, Lord of Luddenham, Imperial College of Science, Technology and Medicine, London, confirmed by electronic mail message (via his secretary Betty Sayers) to Robert Cook-Deegan, National Academy of Sciences, August 10, 1995.

8. Computer Science and Telecommunications Board, National Research Council, *Evolving the High Performance Computing and Communications Initiative to Support the Nation's Information Infrastructure*, 1995.

9. Susan Rosegrant and David R. Lampe, *Route 128: Lessons from Boston's High-Tech Community* (New York: Basic Books, 1992), p. 16.

10. Institute for the Future, *The Future of America's Research-Intensive Industries*, Report R-97, Menlo Park, Calif., 1995.

11. Institute for the Future, *The Future of America's Research-Intensive Industries*, 1995.

12. Alfonso Gambardella, *Science and Innovation: The U.S. Pharmaceutical Industry During the 1980s* (New York: Cambridge University Press, 1995); Rebecca Henderson and Iain Cockburn, *Scale, Scope, and Spillovers: The Determinants of Research Productivity in the Pharmaceutical Industry*, Working Papers Series, Working Paper No. 4466 (Cambridge, Mass.: National Bureau of Economic Research, September 1993); Rebecca Henderson, "The Evolution of Integrative Capability:

Innovation in Cardiovascular Drug Discovery," in *Industrial & Corporate Change*, Vol. 3, No. 3, Winter 1994 (New York: Oxford University Press, 1994), pp. 607-630.

13. Edwin Mansfield, "Academic Research Underlying Industrial Innovations: Sources, Characteristics, and Financing," *The Review of Economics and Statistics*, February 1995, pp. 55-65; Francis Narin and Dominic Olivastro, "Status Report: Linkage Between Technology and Science," *Research Policy* 21: 237-249, 1992; Francis Narin and Richard P. Rozek, "Bibliometric Analysis of U.S. Pharmaceutical Industry Research Performance," *Research Policy* 17: 139-154, 1988.

14. Francis Narin and Dominic Olivastro, "Status Report: Linkage Between Technology and Science," 1992, at p. 248.

15. David Swinbanks, "MITI Clears New Path for Japan's Universities," *Nature* 376 (13 July), 1995; Science and Technology Agency, Japan, *White Paper on Science and Technology, 1995: Fifty Years of Postwar Science and Technology in Japan* (Tokyo: Prime Minister's Office, July 1995).

16. Science and Technology Agency, Japan, *White Paper on Science and Technology*, 1995.

Appendixes

Appendix A
Senate Report Language for the Prospective Study on Allocation of Federal R&D Funding

Excerpt from report language accompanying Public Law 103-733:

Research Report—The [Senate Appropriations] Committee is concerned that at a time when there is such opportunity to understand and cure disease, funding for health research supported by NIH in the next fiscal year is held to below the inflation index for medical research due to budget constraints. Similarly, other Federal research agencies are confronted with constrained resources resulting from the virtual freeze in discretionary outlays. This freeze will make decisions over how to best allocate funding for research and development in the future all the more difficult as research opportunities collide with other governmental responsibilities required for preserving, protecting the health, safety and economic security of our citizen [sic]. These realities have compelled the Committee to consider the composition of the overall Federal Government research and development budget, which currently totals more than $70,000,000,000 a year. In particular, the Committee is concerned whether that research budget is designed to meet new national security concerns, military, economic, and health, that confront our Nation in a post-cold war world. The Committee is concerned, for example, that medical research is not at its optimal level of priority and support relative to its importance to national security.

Because of these new circumstances, the Committee has provided [$750,000] within the Office of the director [of NIH] to commission a study by the National Academy of Sciences and the Institute of Medicine. The study should consider the criteria that should be used in judging the appropriate allocation of funds to research and development activities, the appropriate balance among different types of institutions that conduct such research, and the means of assuring continued objectivity in the allocation process. The academies and Institute should consult with the Office of Science and Technology Policy in planning the framework for the report. The academies and the Institute should submit the report to both the House and Senate Appropriations Committees by December 31, 1995.

Appendix B
Committee and Staff Biographical Information

FRANK PRESS, *Chair*, is the Cecil and Ida Green Senior Fellow at the Carnegie Institution. He served as president of the National Academy of Sciences from 1981 to 1993 and as the president's science adviser during the Carter administration. A geophysicist, he has served on the faculties of Columbia University, California Institute of Technology, and Massachusetts Institute of Technology. He served on the president's Science Advisory Committee during the Kennedy administration and on the presidential Advisory Committee during the Ford administration. He was appointed by president Nixon to the National Science Board of the National Science Foundation and also served on the Lunar and Planetary Missions Board of the National Aeronautics and Space Administration. Among his many honors, Dr. Press received the Japan Prize and the Vannevar Bush Award in 1993 and the National Medal of Science in 1994.

LEW ALLEN, JR., is chairman of the Charles Stark Draper Laboratory, Inc. From 1982 to 1990, he served as vice president of the California Institute of Technology and director of the Jet Propulsion Laboratory. Prior to his association with CalTech and JPL, he served as Air Force chief of staff and as a member of the Joint Chiefs of Staff. Dr. Allen also served as director of the National Security Agency from 1973 to 1977. He is a member of the National Academy of Engineering.

DAVID H. AUSTON is the provost of Rice University. Prior to his appointment at Rice, he was professor of electrical engineering and applied physics and dean of the School of Engineering and Applied Sciences at Columbia University. He also was a member of the technical staff at AT&T Bell Laboratories. He is the recipient of the R.W. Wood Prize from the Optical Society of America, the Quantum Electronics Award from the Institute of Electrical and Electronics Engineers, and the Morris E. Leeds Award. He is a member of both the National Academies of Sciences and Engineering.

FOREST BASKETT is the chief technology officer and senior vice president of research and development at Silicon Graphics. Since joining Silicon Graphics in 1986, Dr. Baskett has led engineering teams in the design of multiprocessing workstations and graphics structures. Before joining Silicon Graphics, Dr. Baskett was the director of Digital Equipment Corporation's Western Research Laboratories, where he was the leader of the Titan Project, designing and building research prototypes of reduced instruction set computing (RISC) systems. Prior to his work with Digital, he spent 11 years as a professor of computer science and electrical engineering at Stanford University. Dr. Baskett is a member of the National Academy of Engineering.

BARRY R. BLOOM is the Weinstock Professor of Microbiology and Immunology at the Albert Einstein College of Medicine and an investigator for the Howard Hughes Medical Institute. He has been a consultant to the White House in international health policy, chaired committees at the World Health Organization, and served on National Research Council committees. He was president of the American Association of Immunologists, and of the Federation of American Societies for Experimental Biology. He received the first Bristol Myers Squibb Award for Distinguished Research in Infectious Diseases and the Mayor of New York's Award for Excellence in Science and Technology. He is a member of the National Academy of Sciences and of the American Academy of Arts and Sciences, and is a councillor of the Institute of Medicine.

DANIEL J. EVANS, chairman of Daniel J. Evans Associates, is a former governor, United States senator, and state house of representatives member of Washington State. He was also president of the Evergreen State College in Washington. Mr. Evans currently serves on the boards or advisory committees of a number of philanthropic and business associations, including the Board of Regents of the University of Washington. He was the chair of the Panel on Policy Implications for Global Warming of the Committee on Science, Engineering, and Public Policy.

BARUCH FISCHHOFF is a professor of social and decision sciences and of engineering and public policy at Carnegie Mellon University. He is the recipient of the American Psychological Association's Early Career Awards for distinguished scientific contributions to psychology and for contributions to psychology in the public interest. He is a fellow of the Society for Risk Analysis, as well as a recipient of its Distinguished Achievement Award. His current research includes risk communication, adolescent decision making, evaluation of environmental damage, and insurance-related behavior. He is a member of the Institute of Medicine.

MARYE ANNE FOX is the vice chair of the National Science Board. She is vice president for research at the University of Texas and the M. June and J. Virgil Waggoner Regents Chair in Chemistry. Dr. Fox was the recipient of the Garvan Medal and the Arthur C. Cope Scholar Award from the American Chemical Society. She is a member of the National Academy of Sciences and of the American Academy of Arts and Sciences, and a former member of the National Research Council's Commission on Physical Sciences, Mathematics, and Applications.

SHIRLEY A. JACKSON, professor of physics at Rutgers University, was recently appointed by President Clinton as chair of the U.S. Nuclear Regulatory Commission. A theoretical physicist, she spent 15 years on the technical staff at AT&T Bell Laboratories. She is a fellow of the American Academy of Arts and Sciences and at the American Physical Society and is a member of the MIT Corporation. She is a member of the National Research Council's Commission on Physical Sciences, Mathematics, and Applications. Dr. Jackson resigned from the committee on July 12, 1995, to become the chair of the U.S. Nuclear Regulatory Commission.

ROBERT I. LEVY is president of Wyeth-Ayerst Research. He previously served as president of the Sandoz Research Institute and professor of medicine at Columbia University. His other appointments include vice president for health sciences at Columbia and Tufts Universities; dean, Tufts University School of Medicine, and director of the National Heart, Lung, and Blood Institute. In 1980, he received the Albert Lasker Special Public Health Award and in 1988 was presented with the Humana Heart Foundation Award for outstanding and long-term contributions to the field of cardiology and cardiovascular medicine. He is a member of the Institute of

Medicine. Dr. Levy resigned from the committee on March 22, 1995, due to schedule conflicts.

RICHARD J. MAHONEY recently retired as the chairman and chief executive officer of Monsanto Company. He joined Monsanto in 1962 as a product development specialist and subsequently held various marketing, technical service, and new product development positions in Plastic Products, Agriculture, and International Operations. He was elected president in 1979 and named chief executive officer in 1983. He is a director of the Metropolitan Life Insurance Company and of the Union Pacific Corporation, as well as a member of the advisory committee and/or board of numerous philanthropic, educational, and business associations.

STEVEN L. McKNIGHT is research director for Tularik, Inc. He also served as a staff member of the Department of Embryology at the Carnegie Institution and as an investigator for the Howard Hughes Medical Institute. He is a lecturer and a recipient of the Eli Lilly Award, the AAAS Newcomb Cleveland Award, and the National Academy of Sciences Award in Molecular Biology (the Monsanto Award). He is a member of the National Academy of Sciences.

MARCIA K. McNUTT is the Griswold Professor of Geophysics at the Massachusetts Institute of Technology. Prior to coming to MIT, she was a geophysicist for the U.S. Geological Survey. She is the recipient of the Macelevane Award of the American Geophysical Union. Dr. McNutt has served on the National Research Council's Committee on Earth Sciences and has been an active participant in the National Academy of Sciences' annual symposia on the Frontiers of Science.

PAUL M. ROMER is professor of economics at the University of California at Berkeley and a research associate for the National Bureau of Economic Research. He previously held teaching positions at the University of Chicago and University of Rochester. Dr. Romer was the recipient of a Sloan Foundation fellowship.

LUIS SEQUEIRA is the J.C. Walker Professor of Plant Pathology at the University of Wisconsin in Madison. A plant pathologist, he served as the director of the Coto Research Station in Costa Rica and on the faculty of North Carolina State University. His current research interests include soil microbiology, root diseases, plant growth regulators, and the physiology of parasitism. He is a member of the National Academy of Sciences.

HAROLD T. SHAPIRO is president of Princeton University, where he also is professor of economics and public affairs. Prior to coming to Princeton, he served on the faculty of the University of Michigan for 24 years as professor of economics and public policy; he was president from 1980 to 1988. His professional activities include memberships on the Conference Board, Inc. and the Bretton Woods Committee, as well as on the boards of many other corporations and institutions. From 1990 to 1992, Dr. Shapiro served as a member of President Bush's Council of Advisors on Science and Technology. He is a member of the Institute of Medicine and chaired the Institute's Committee on Employer-Based Health Benefits.

H. GUYFORD STEVER, a corporate director, scientist, and engineer, served as White House science and technology adviser to President Ford, director of the Office of Science and Technology Policy, and director of the National Science Foundation from 1973 to 1976. Prior to his government service, he was president of the Carnegie-Mellon University from 1965 to 1972 and professor of aeronautics and astronautics at the Massachusetts Institute of Technology for 20 years. Dr. Stever has served as the National Academy of Engineering's foreign secretary and as chairman of the National Research Council's Committee on Space and of the Panel of Technical Evaluation of NASA's Proposed Redesign of the Space Shuttle Solid Rocket Booster. He is a member of both the National Academies of Sciences and Engineering.

JOHN P. WHITE was, until recently, the director for the Center for Business and Government at the John F. Kennedy School of Government at Harvard University. He took over the program following his active involvement in both the Clinton and Perot 1992 presidential campaigns. He was general manager of the Integration and Systems Products Division and vice president of Eastman Kodak Company and the chief executive officer and chairman of the board of Interactive Systems Corporation from 1981 until it was sold to Eastman Kodak in 1988. Previously, he served in the federal government as the deputy director of the Office of Management and Budget and as the assistant secretary of defense, manpower, reserve affairs and logistics. Mr. White resigned from the committee on June 22, 1995, to become deputy secretary of defense.

Staff

NORMAN METZGER is executive director of the Commission on Physical Sciences, Mathematics, and Applications, one of the major program units of the National Research Council. Prior to assuming this position in 1990, he was deputy executive officer of the National Research Council. He has been with the Council since 1975, and before that held positions with the American Association for the Advancement of Science, the American Chemical Society, and the Sloan-Kettering Institute. He holds a degree in chemistry and was a Sloan-Rockefeller fellow in advanced science writing. He has written books on chemical research and energy supply and demand, as well as numerous articles on science and technology for *Popular Science, New Science,* and other publications.

ROBERT M. COOK-DEEGAN is a physician-molecular biologist, formerly director of the Institute of Medicine's Division of Biobehavioral Sciences and Mental Disorders. He has supervised eight major projects and numerous smaller efforts since joining the Institute of Medicine in early 1991. He previously directed several studies for the Office of Technology Assessment, where he was a senior associate, and was acting director of the Biomedical Ethics Advisory Committee, an analytical support agency of the U.S. Congress in 1988 and 1989. He has authored over 100 articles on various topics and recently published a book, *The Gene Wars: Science, Politics and the Human Genome*, with W.W. Norton & Co., New York (1994). He obtained his bachelor's degree in chemistry from Harvard College in 1975 and his MD from the University of Colorado in 1979.

CHRISTOPHER T. HILL is professor of public policy and technology in the Institute of Public Policy at George Mason University. Before joining George Mason University, Dr. Hill was at the RAND Critical Technologies Institute. He also served as executive director of the Manufacturing Forum at the National Academy of Engineering and the National Academy of Sciences. A chemical engineer, he served as senior specialist in science and technology policy at the Congressional Research Service. He has taught at Washington University and Massachusetts Institute of Technology and worked at the Office of Technology Assessment and Uniroyal.

MICHAEL G.H. McGEARY is a political scientist who directed the staff work for 10 major reports by various units of the National Academy of Sciences (Institute of Medicine, Commission on Behavioral and Social Sciences and Education, Office of Scientific and Engineering Personnel, and Committee on Science, Engineering, and Public Policy) between 1981 and 1995. He did his graduate work at Massachusetts Institute of Technology and, prior to coming to the National Academy of Sciences, he worked at the National Academy of Public Administration and taught at Wellesley College. Currently he is a consultant and is coauthoring a book on U.S. science and technology policy.

JULIE M. ESANU is a research assistant with the Commission on Physical Sciences, Mathematics, and Applications. She works primarily on scientific and technical data information issues. She received her bachelor's degree in political science from the George Washington University in 1989.

DANIELLE DEHMLER is a project assistant to this study. She received her bachelor's degree in legal studies from the State University of New York at Buffalo in 1993.

Appendix C
Acknowledgments

The Committee on Criteria for Federal Support of Research and Development is very grateful to the many individuals who played a significant role in the completion of this study. The committee met four times for 10 days, and extends its gratitude to the following individuals who appeared before the full committee to provide background information and discuss pertinent issues: Marvin Cassman, acting director of the National Institute of General Medical Sciences, National Institutes of Health; Ruth Kirschstein, deputy director, National Institutes of Health; Judy Vaitukaitis, director, National Center for Research Resources, National Institutes of Health; Harold Varmus, director, National Institutes of Health; France Cordova, chief scientist, National Aeronautics and Space Administration; Daniel Goldin, administrator, National Aeronautics and Space Administration; James Decker, deputy director, Office of Energy Research, Department of Energy; Martha Krebs, director, Office of Energy Research, Department of Energy; Essex Finney, associate administrator, Agricultural Research Program, Department of Agriculture; David Goldston, former project director, Council on Competitiveness; Mary Good, undersecretary for technology, Department of Commerce; Senator Tom Harkin, (D-IO); Robert Hermann, senior vice president for science and technology, United Technologies Corporation; Robert Huggett, assistant administrator for research and development, Environmental Protection Agency; Anita Jones, director, Defense Research and Engineering, Department of Defense; Neal Lane, director, National Science Foundation; Anne Petersen, deputy director, National Science Foundation; Kathleen Peroff, deputy associate director, Energy and Science Division, Office of Management and Budget.

The committee also extends its thanks to the following members of Congress and congressional staff who provided background and additional information to the chair and staff: Senator Christopher Bond (R-MO), chair, Senate Appropriations Subcommittee on Housing and Urban Development, Veterans Affairs, and Independent Agencies; Congressman Robert Walker (R-WI), chair, House Science Committee, and vice-chair, House Budget Committee; and the staff of the House Science Committee, including Anne Marcantognini, deputy chief of staff; Michael Rodemeyer, chief democratic counsel; Deirdre Stach, budget analyst; Ed McGaffigan, senior political adviser for budget, defense, foreign relations, and veterans for Senator Jeff Bingaman; Craig Higgins, majority clerk for the Senate Appropriations Subcommittee on Labor, Health and Human Services, and Education; and Ed Long, former majority clerk for the Senate Appropriations Subcommittee on Labor, Health and Human Services, and Education.

The committee is grateful for the efforts of the following members of the Office of Science and Technology Policy: John H. Gibbons, assistant to the President for science and technology; MRC Greenwood, former associate director for science; Lionel Johns, associate director for technology; Jane Wales, associate director for national security and international affairs; Robert Watson, associate director for environment; Catherine Woteki, acting associate director for science; and Angela Phillips Diaz, executive secretary.

For their assistance in data gathering, preparation, and consultation the committee extends its thanks to the following individuals: Harriet Dustan, member, Institute of Medicine; Ed Roccella, coordinator, National High Blood Pressure Education Program; Rebecca Henderson, associate professor of management, Massachusetts Institute of Technology; Russell Herndon and Robert Tuohy, Defense Research and Engineering, Department of Defense; Harvey Brooks, professor of technology and public policy, John F. Kennedy School of Government, Harvard University; Bruce Fonoroff, Army Research Laboratory; Howard Garrison, Federation of American Societies for Experimental Biology; Robert Levy, president, Wyeth-Ayerst Research; Stanley Trice, analyst, Defense Research and Engineering, Department of Defense; Jane Bortnick Griffiths, acting chief, Science and Technology Division, Congressional Research Service; Genevieve Knezo, Congressional Research Service; Richard Rowberg, Smithsonian Institution; Kei Koizumi, Kathie Gramp, and Al Teich, American Association for the Advancement of Science; Robert Smith; David Guston, Rutgers University; Philip Smith; Marvin Ebel, Council on Governmental Relations; John Jankowski, Ronald Meeks, and Linda Parker, National Science Foundation; David Kingsbury, director, Genome Database, John Hopkins University; Michael Crow, Columbia University; Ann Markusen, Rutgers University; Donald Stokes, Center for Advanced Study; Richard Nelson, Columbia University; Kitty Gilman, National Science and Technology Council; Donna Fossum and Tim Webb, Critical Technologies Institute, RAND; J. Michael Bishop, University of California, San Francisco; Marjory Blumenthal, director, Computer Science and Telecommunications Board, National Research Council; A. Michael Spence, chair, Committee on Science, Technology, and Economic Policy, National Academy of Sciences; Bob Bayer, Department of Defense; Bob Meisner.

To gather views from a broad range of interests, the committee organized outreach sessions to help frame its observations and recommendations. The first session was held at Stanford University on February 21, 1995. Another was held at the University of Texas at Austin on April 7. To continue the dialogue the committee also took advantage of previously scheduled meetings such as the American Association for the Advancement of Science annual meeting, the Sigma Xi Forum, and the annual meetings of the National Academies of Sciences and Engineering. The committee is grateful to all those who attended its outreach sessions and to the following individuals for their assistance with its outreach efforts: Gerhard Casper, president, Stanford University; Charles Kruger, vice provost for research and policy, Stanford University; Kathy Eslinger, executive assistant to the vice provost for research and policy, Stanford University; and Nancy Mallory and Susie Pruett, assistants to the vice president of research and development, University of Texas at Austin.

Finally, the committee would like to recognize the special contributions of both the National Research Council staff and the independent consultants who served on the study: Norman Metzger, executive director of the Commission on Physical Sciences, Mathematics, and Applications, who served as the study director; Robert Cook-Deegan of the Institute of Medicine, who served as the senior program officer; Michael McGeary, Christopher T. Hill, and Patrick Young, who served as consultants; Julie Esanu, for the program and research assistance provided to the committee; Danielle Dehmler, for the staff support for the committee and for her work in preparing the final manuscript; and Susan Maurizi, who edited the final manuscript.

Appendix D
List of Commissioned Background Papers

On the Allocation of Federal R&D
Christopher T. Hill, Professor of Public Policy and Technology
George Mason University

This paper provides a broad overview of the present-day allocations of federal research and development (R&D) funds among diverse national purposes, performers, and sponsoring federal agencies. It sets out a number of general considerations in the formulation of federal R&D policies over the past 5 decades, and it describes and analyzes the historical processes through which the portfolio of federal R&D has been shaped.

Previous Analyses of the U.S. R&D Allocation Process
Robert M. Cook-Deegan, Senior Program Officer
National Research Council

This document reviews some major reports on U.S. federal science policy relevant to the committee's task. Its purpose is to review what others have said when confronted with tasks similar to the committee's mandate, or when given different tasks that required them to confront similar choices about R&D spending that must be made at the presidential or congressional level.

Where Does the Federal Dollar for Basic Research Go?
Michael G.H. McGeary, Consultant

This paper describes the composition of the federal basic research budget and presents background information on federally funded basic research—who funds basic research and why, what mechanisms are used to provide support, who conducts basic research and how, and how performers spend the dollars.

Survey of Reports on Federal Laboratories
Robert M. Cook-Deegan, Senior Program Officer
National Research Council

This background paper summarizes the recent reviews of federal laboratories, both those operated by government and those conducted by contractors (whether corporate or university).

NOTE: Copies of all commissioned papers are available upon request from the Commission on Physical Sciences, Mathematics, and Applications, National Research Council, 2101 Constitution Avenue, NW, Washington, DC 20418.

Appendix E
Acronyms

AAAS	American Association for the Advancement of Science
ACE	Angiotensin converting enzyme
AIDS	Acquired immune deficiency syndrome
ARPA	Advanced Research Projects Agency
ATP	Advanced Technology Program
CDC	Centers for Disease Control and Prevention
COSEPUP	Committee on Science, Engineering, and Public Policy
CRADA	Cooperative research and development agreement
DHHS	Department of Health and Human Services
DNA	Deoxyribonucleic acid
DOD	Department of Defense
DOE	Department of Energy
DVA	Department of Veterans Affairs
EPA	Environmental Protection Agency
FDA	Food and Drug Administration
FFRDC	Federally funded research and development center
FS&T[1]	Federal science and technology
GDP	Gross domestic product
GOCO	Government-owned, contractor-operated laboratory
GOGO	Government-owned, government-operated laboratory
GPRA	Government Performance and Results Act of 1993
GPS	Global positioning system
HIV	Human immunodeficiency virus
HUD	Department of Housing and Urban Development
MEP	Manufacturing Extension Partnerships program
NASA	National Aeronautics and Space Administration
NIH	National Institutes of Health
NIST	National Institute of Standards and Technology
NOAA	National Oceanic and Atmospheric Administration
NSF	National Science Foundation

---

[1] A term introduced by the committee to denote the science and technology base that underlies the broader national R&D enterprise, as well as the pool of federal funding that supports that base.

NSTC	National Science and Technology Council
ODDR&E	Office of the Director of Defense Research and Engineering
OMB	Office of Management and Budget
ONR	Office of Naval Research
OSTP	Office of Science and Technology Policy
R&D	Research and development
RISC	Reduced instruction set computing
TRP	Technology Reinvestment Program
USAID	U.S. Agency for International Development
USDA	U.S. Department of Agriculture
USGS	U.S. Geological Survey
VLSI	Very large scale integrated circuit program